AF540140

An Introduction to Analytical Chemistry

S.A. IQBAL
Dept. of Chemistry
Saifia P.G. College of
Science and Education
Bhopal, INDIA

M. SATAKE
Faculty of Engineering
Fukui University
Fukui 910, Japan

Consulting Editors

Y. MIDO
Dept. of Chemistry
Kobe University
Kobe 657, Japan

M.S. SETHI
A.R.S.D. College
University of Delhi
New Delhi-INDIA

Discovery Publishing House
New Delhi (INDIA)

First Published – 1999

Reprinted – 2025

ISBN: 978-81-7141-244-0

© Authors

An Introduction to Analytical Chemistry

Published by:

DISCOVERY PUBLISHING HOUSE PVT. LT

4383/4B, Ansari Road, Darya Ganj

New Delhi-110 002 (India)

Phone: +91-11-23279245; 23253475; 43596065

Mobile: +91 9811179893 / +91 9871656464

E-mail: discoverybooksindia@gmail.com

orderdphbooks@gmail.com

namitwasan9@gmail.com

web: www.discoverypublishinggroup.com

Printed at:

Infinity Imaging Systems

Delhi (INDIA)

Preface

The teaching of Chemistry at the introductory stage becomes each day a more challenging task as the subject matter becomes more diverse and more complex. These challenges have evoked a series of responses – the present set of introductory chemistry monographs is one such. The teaching of chemistry recognises a number of problems that confront those who select text books. In order to overcome these problems, this volume "**An Introduction to Analytical Chemistry**" – one of about fifty in the Chemistry Monograph Series – is introduced. Each volume is independent of the others deals with one of Chemistry topics and constitutes a complete entity. Each volume is more comprehensive than can be possible in a single volume text. It is intended to provide a range of topics to cover most undergraduate and chemistry main courses of study. These volumes can be used to enrich the more conventional courses of study.

Suggestions for improvement are welcome and shall be gratefully acknowledged.

Authors

Contents

CHAPTER 1

Introduction

THE IMPORTANCE OF ANALYTICAL CHEMISTRY

Chemical analysis holds an important position in our modern society. A knowledge of analytical procedures has become essential in practically all the arts and sciences as well as the technical operations of industry. The discovery of new medicinals, new fabrics, new alloys, and other materials would not be possible without the services of the analytical chemist. Analytical chemistry is such an important part of our industrial structure and advancing science that without it out present day economy could not exist.

In the early days of chemistry, analysis was almost synonymous with chemistry itself. Robert Boyle (1626-91) first introduced the word "analysis" to apply to the identification of individual substances in the presence of each other. From his interest in analysis it was recognized that advancement in chemical knowledge could come only from the experimental method of the laboratory. Berzelius (1779-1848) was a master in the field of chemical analysis. He developed many new experimental methods which included the separation of the metal constituents of a number of complex minerals. In 1818 he published a table of atomic weights, based on analytical procedures, which was a model of quantitative accuracy for many years. Heinrich Rose (1795-1864), a pupil of Berzelius, published one of the first texts to deal exclusively with analytical chemistry : "Ausfuhrliches Handbuch der Analytischen Chimie" (Detailed Textbook of Analytical Chemistry).

The last twenty years have witnessed an expanding interest in the problems and methods of chemical analysis. This has been the result of many causes : the great increase in industrial production, the great number and variety of industrial products, the more careful control that is required in the manufacture of these products, and the necessity for speed in the securing of analytical information no longer possible by the classical methods of chemical analysis. Many companies that in the past were able to succeed without analyzing their raw materials or their products, have recently found that it is now essential to make analyses in order to maintain quality and uniformity. The purchaser demands assurance that he can obtain, in the future, materials of the quality and uniformity that he now secures. Another element in the expanding analytical picture has been the demand caused by recent wars for substances urgently needed for military purposes as well as substitutes for scarce materials. The atomic energy project alone has required a large amount of research in analytical methods.

There has been a vast increase in the number of analyses that are needed in the daily problems of engineering, metallurgy, manufacturing, and many other fields. It is often necessary to determine traces of elements so small that chemical tests fail and sensitive physical instruments, such as the spectrograph, the mass spectrometer, and the Geiger counter, must be used. The properties of our common metals, aluminum, iron, copper, zinc, lead, etc., are tremendously influenced by the presence of other elements either added intentionally or known as impurities. It is a vital matter, therefore, that the composition of these metals be known accurately. Manufacturers have established large expensive laboratories, with a wealth of equipment and with large and versatile staffs to operate them, for the sole purpose of analyzing rapidly and accurately their raw materials and products. Between five and six million analyses are made in a single year by one large company.

Agriculture has been aided with increased yields and enhanced soil fertility through accurate chemical analysis by adjusting the composition of fertilizers and by effective use of insecticides and pesticides. In particular, the presence of proper amounts of trace elements such as zinc, iron, manganese, molybdenum, cobalt, and boron which are essential to plant growth is ensured by good chemical control. The cause of the disease of cattle in certain regions of western United States was unknown until chemical analysis revealed the presence of selenium in the vegetation. Similarly, a disease of sheep was traced to the absence of cobalt in soil and plants.

Geologists have been interested in the composition of meteorites and in the age of rocks which they have determined by the amount of uranium, thorium, and pure isotopes of lead which they contain. Geological prospecting has recently been aided by chemical analysis of the ash of vegetation.

Advances in medicine and public health would not be possible without chemical analysis. An enormous number of chemical analyses is made daily in the laboratories of hospitals for purposes of diagnosis and treatment. Analyses of blood, urine, etc., are absolutely essential in order that proper treatment may be prescribed. An excess or deficiency of sugar, calcium, potassium, iodine, or other substances may be the cause of a pathological condition. Poisons may be identified by analysis, and often their source is indicated. The importance of fluoride in potable waters and its influence on dental caries was discovered only by careful analysis of these waters; its addition to municipal water supplies is possible only with careful analytical control.

The historian and the archaeologist have obtained valuable information from chemical analyses of soil, building materials, various objects of art, and coins recovered in excavations of various sites. Certain objects of art, when associated with ancient coins, can be dated roughly from the chemical composition of the coins which underwent progressive debasement that paralleled increasing economic difficulties that occurred with the decline of the empire.

The progress of many a research problem has been retarded by lack of a suitable analytical method or lack of analytical data. It has been said that a good analyst furnishes "the eyes through which research sees its way forward." The work of the analytical chemist, although often unnoticed and rarely dramatized, contributes in a very important way to research, general welfare, and progress.

Training in analytical chemistry serves as a sound basis for studies in other branches of chemistry and other sciences. The laboratory work of physical chemistry requires the quantitative skills that are developed in the analytical course. Work in biology, geology, and physics may be accomplished at a higher level of understanding and ability with a background of the principles and techniques acquired in a course in quantitative analysis. Medical and dental schools recognize the importance of quantitative analysis by requiring it as a prerequisite for admission to the study of medicine.

THE SCOPE OF QUANTITATIVE ANALYSIS

Qualitative analysis is concerned with the identification and separation of chemical substances and the chemical principles on which such procedures are based. Several schemes of analysis are in general use, the most common being the so-called sulfide method of separation.

Quantitative analysis is concerned with the determination of the amount of a chemical substance present wither alone or in a simple or complex mixture of other substances. Unlike qualitative analysis, there is no general scheme of separation but rather a scheme of experimental techniques that find general use after the isolation of the particular substance to be determined. In certain physico-chemical procedures the isolation may not be necessary. The literature of quantitative analysis is extensive and there are many scientific journals devoted exclusively to the subject. A selection of some of these sources of information is given in the Appendix.

Quantitative analytical methods may be broadly divided into two general groups.

Group I consists of those methods in which the final measurement of the substance sought is made by direct of indirect measurements of volume and weight after proper treatment of a measured portion of the material to be analyzed. The two most important classifications in this group are gravimetric methods and titration methods. The latter are often referred to as volumetric or titrimetric analysis.

Gravimetric Methods

In these methods the analysis is carried out by a series of weighing operations. The most usual is the isolation of the substance sought by direct precipitation of the substance, or a suitable compound of it, with subsequent purification before weighing. In some methods the substance is deposited electrically on a suitable electrode and then weighted. Another method involves the evolution of a gas. The gas or vapor may be evolved by heat or a chemical reaction and its weight determined by the gain in weight of the vessel in which absorption takes place. Alternatively, the loss in weight of the original substance may be used.

The most commonly used gravimetric methods for an introduction to quantitative analysis are the determination of chloride and of sulfate in water-soluble samples in the absence of interfering substances. The chloride is determined by precipitation as silver chloride and the sulfate by precipitation as barium sulfate. The favored determination by electroanalysis is the deposition of copper on a platinum cathode. The usual gravimetric separations include the analysis of brass and limestone. In the former, tin, lead, copper, and zinc are separated from each other and determined in suitable insoluble forms. The latter involves the separation and determination of silica, hydrous oxides, calcium, and magnesium. Loss on ignition is also sometimes included.

Titration Methods

In these methods the substance sought is determined by a careful measurement of the volume of a solution of known concentration required to react with the solution to be analyzed. Titration methods fall into four general groups. These are : (1) neutralization reactions involving the determination of acids and bases ; (2) oxidation-reduction reactions involving the determination of reducing agents such as Fe^{II}, $C_2O_4^{2-}$, Sn^{II}, As^{III}, and oxidizing agents such as Cu^{II}, I_2, $Cr_2O_7^{2-}$ and MnO_4^- ; (3) precipitation reactions which are usually limited in an introductory course to Cl^- ; (4) complex-formation reactions which involve the formation of a stable complex ion or a slightly dissociated molecule (other than water).

The volume of reagent required for the completion of the reaction in these methods is usually measured by a visible color change that occurs in the reaction itself or is produced by the presence of a chemical indicator. The completion of the reaction may also be measured by some physicochemical change such as a change in potential, electrical conductance, absorption of radiation, change in temperature, and many others. Most of these latter methods are used only in advanced courses.

Group II consists of those methods in which the final measurement is made upon the system as a whole. Almost any physical measurement may be used in this group to give the desired analytical result. Comparison with mixtures of known composition is necessary. The procedures are so varied that only a few illustrative examples are given here. The measurement of specific gravity serves as the basis for the analysis of such binary mixtures as alcohol and water. Thermal measurements such as melting points give information on the purity or quantitative composition of substances or binary mixtures. Electrical measurements of conductance, electromotive force, dielectric constant, etc., are used to determine the composition of a mixture from a single physical measurement. Optical methods include ; emission of spectra, absorption of radiation, diffraction, fluorescence, refraction, and many others.

The methods outlined in Group I are those studied in the introductory course in quantitative analysis. A suitable selection from these gives the student familiarity with fundamental principles and laboratory techniques in the quantitative handling of materials. The methods of Group II assume the background of those in Group I and are studied in advanced or specialized courses. Methods selected from Group I are stressed in introductory courses because the principles and the techniques are essential in connection with other courses in chemistry and in many fields of scientific study such as biology, geology, physics, and medicine, as well as in the practical application of analytical chemistry.

CHAPTER 2

Introduction to Laboratory Work

Success in the laboratory work of quantitative analysis is dependent on adherence to certain well-recognized rules and regulations.

Neatness is essential. Glassware should be kept clean, apparatus should be kept in an orderly fashion in the locker, and the working space on the bench should be carefully wiped clean of all puddles of liquid and spilled solids before any operation is started and after it is completed. Sideshelf reagents should be returned to their proper place after use, and any spilled should be promptly cleaned up.

Advance knowledge and planning before a procedure is started will give efficient use of the time available. The directions for all procedures should be read in advance to be sure that the necessary apparatus and reagents are available before the beginning of the laboratory period. Much time may be wasted by failing to recognize that certain operations such as the drying of samples or precipitates, and evaporations are time-consuming. They cannot be hurried, but while they are taking place certain other procedures can be carried out. For example, a crucible may be ignited and weighed while a precipitate is digesting or drying.

Each operation should be performed with extreme care. Some operations may be quite time-consuming, but speed develops as one gains in experience. For beginners, it is only natural that difficulties will arise which may require the repetition of a lengthy procedure. A student who encounters no difficulties is decidedly in the minority.

LABORATORY EQUIPMENT AND MATERIALS

The Notebook

One of the objectives of a course in quantitative analysis is training in the proper recording of scientific data. This requires a permanently-bound notebook. A convenient size for this notebook is about 5 ½ × 8 inches. It may open at the side or at the top. The latter type is very useful if the top cover folds under the bottom cover. This results in a convenient pad that occupies a minimum of desk space. Certain types are made with detachable pages with carbon paper inserted. The advantage of this is that carbon copies may be turned in as report sheets so that inspection of student's work may be made without requiring the whole notebook. Also, if the notebook is lost, carbon copies of the data remain available.

The subject of the procedure and the date should first be written down. A series of buret readings or weighings should be neatly recorded in the form of a table and entries made *at the time an observation is taken.* Erasures should never be made, but figures that are recorded in error should be indicated as discarded. The method of calculation of results should be clearly indicated on a page or section of a page of the notebook. Where logarithms are required they should be included. The meaning of each figure in the data or calculations should be clearly labelled so that anyone who looks at the notes can understand them without question.

The practice of recording data on scraps of paper, envelopes, and the backs of old letters, should not be tolerated. There should be one and only one notebook, and all data should be recorded in it at the time of observation. Workers in industrial laboratories are required to record all data in a permanent notebook. These notebooks are often filed away in a safe place at the end of the day. When filled with data, they are placed in storage for future reference. They are an invaluable, firsthand reference to the work originally performed.

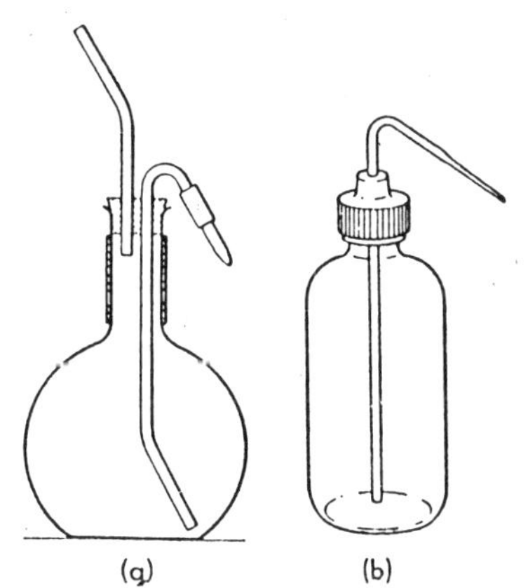

Fig. 2-1. Wash bottles. (a) Common types. (b) Polyethylene type. The flow of liquid is controlled by squeezing the flexible bottle.

Desk Equipment. In addition to the standard items of glassware and other apparatus, certain other items are required in quantitative work. These include : a wash bottle, a desiccator, porcelain crucibles, filtering crucibles, a filter flask and crucible holder, weighing bottles, funnels, stirring rods and policemen, burets, pipets and volumetric flasks.

A wash bottle is a necessary piece of equipment. It is used to wash precipitates and rinse off reagents from glassware. It is prepared by bending suitable pieces of glass tubing which are inserted in a rubber stopper fitted to a 1 litre or 500 ml. flask (see Fig. 2-1a). For handling hot liquids the neck may be wrapped with a thin asbestos sheet or heavy twine. Also for this purpose flasks may be purchased with a wickerwork protection around the neck. Wash bottles of polyethylene are now available (see Fig. 2.1b). The flexible, unbreakable plastic gives easy control of a stream of wash liquid by simply squeezing the bottle.

The *desiccator* (Fig. 2.2), is the most expensive piece of desk equipment, hence care should be used in handling it. Its purpose is to provide a storage place of low moisture content for dried sample and crucibles. It is used to carry samples from the laboratory to the balance room. The bottom is charged with fresh drying material of suitable nature. Anhydrous calcium chloride is commonly used. Other desiccants include concentrated sulfuric acid, phosphorous pentoxide, anhydrous magnesium perchlorate (Dehydrite) and calcium sulphate hemihydrate (Drierite). A wire gauze is often placed above the drying agent and on this rests a tile plate with openings for crucibles or weighing bottles. A suitable lubricant is applied on the unit is thus sealed from direct contact with the atmosphere. The lid is always opened with a sliding motion.

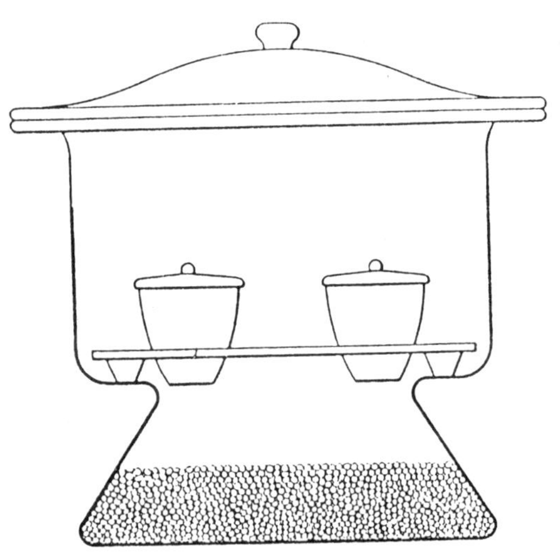

Fig. 2.2. Desiccator and crucibles.

Porcelain crucibles are used to ignite precipitates that are collected on filter paper. Before use the crucible must be marked for identification. A neat way is to make a series of small nicks on the edge with a file. Another method commonly used it to mark lightly the outer surface with a special ink or pencil, dry, and then fuse the mark into the glaze by igniting the crucible to red heat. If three or more crucibles are available it is convenient to make a rough preliminary weighing and then number then in the order of increasing weight. They are then ignited under the same conditions that will be subsequently used, cooled, weighed, and stored in the desiccator.

Filtering crucibles are used in filtering precipitates by suction where the precipitate may be either dried or ignited at a moderate heat. In the latter case, the crucible is placed inside an ordinary porcelain crucible during the ignition process. Filtering crucibles are of three general types. The Gooch type (Fig. 2.3a), the one commonly used in student work, is a porcelain crucible with a large number of small holes in the bottom. Before use, these holes are covered with a mat of especially-treated asbestos fibers, and the crucible is dried and weighed. Glass filter discs may also be used in place of asbestos. The preparation and use of the Gooch crucible is discussed in the following text.

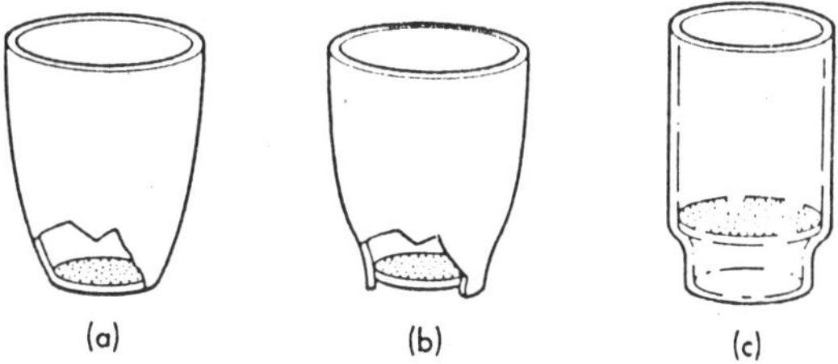

Fig. 2.3. Filtering crucibles. (a) Gooch crucible. The perforated bottom's covered with an asbestos filter mat as described on page 17. (b) Selas or Coors crucible. A porcelain filter bed is sintered into the bottom. (c) Glass filtering crucible. Glass particles sintered together give a filtering medium of any desired porosity.

A second type, called the Selas or Coors crucible (Fig. 2.3b), is made of porcelain and has a fused-in, porous bottom. For general use the pore diameter is of the order of 9-10 microns. These crucibles are more fragile than the Gooch type. If not subjected to a too sudden thermal or mechanical shock, they may be heated to temperatures up to 1200° C.

A third type of filtering crucible is made entirely of glass with a fritted glass disc for a bottom. See Fig. 2.3c. They come in various size and porosities. They may be heated to 500° C. but since they are subject to thermal shock, they must be brought to this maximum temperature by gradual application of heat followed by gradual cooling. Crucibles of this type may be used only with precipitates that may be dissolved off the fritted glass. They, as well as the Selas-type crucibles, should never be used with strongly alkaline materials.

Crucible holders are used to hold a filtering crucible during the process of filtration by suction. There are a number of types in general use. These include : (a) the Sargent, (b) the Bailey, (c) the Walter and (d) the filter tube type. They are illustrated in Fig. 2.4. Their use is given on page 16 in the discussion of filtration by suction.

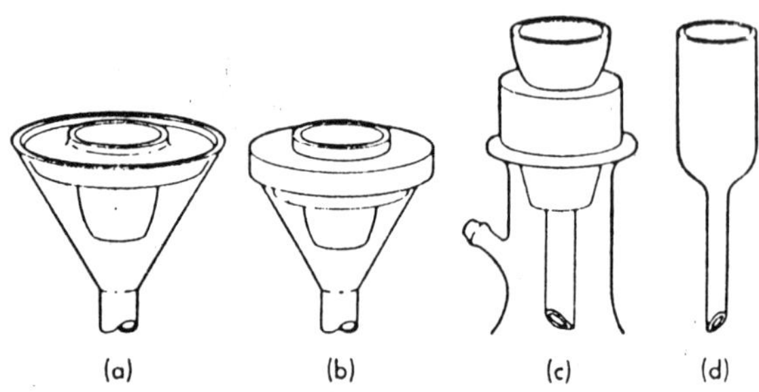

Fig. 2.4. Crucible holders. (a) Sargent type. A circular piece of rubber fits in the top of a funnel. (b) Baitey type. A circular piece of rubber fits on top of a funnel. (c) Walter type. A pear shaped rubber holder with a glass stem fitted to the top of a suction flask. (d) Glass filter tube type. A piece of gum rubber tubing around the top holds the crucible.

Weighing bottles are small, glass-stoppered vessels that are used for the storing of samples prepared for analysis. There are two general types : (a) those with ground, inside-fitting stoppers, and (b) those with outside caps. See Fig 2.5. Identification numbers are etched on each bottle. The ground areas may be used for pencil markings for identification purposes.

Glass funnels are used in filtering operations involving the use of filter paper. To speed up the usually slow filtration process, some funnels (Kimble brand) are molded to an angle of 58° while others (Pyrex brand) are pressed from blanks accurate to 60°. These latter come fluted and plain. Each type claims special merits.

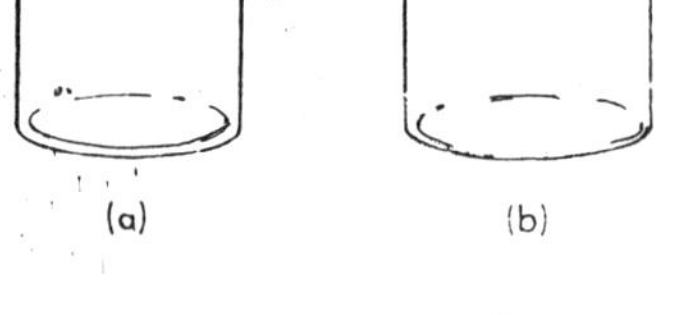

Fig. 2.5. Weighing bottles. (a) Ground, inside-fitting stopper type. (b) Ground, outside-cop type.

Stirring rods of various lengths are required ; they should be fire-polished at both ends after cutting to such lengths that about 2 cm. of the rods project above the lip of the beaker or other vessel with which the rod is to be used. One of the rods is used to attach the rubber *policeman,* a solid piece of soft rubber used to loosen precipitates that may adhere to the inner walls of a beaker. See Fig. 2.6.

Glass hooks are made by bending short pieces of glass rod in the form of an angle. These are used to aid evaporations. Three are hung on the edge of a beaker before the vessel is covered with a watch glass. An open space is thus provided around the top while the contents remain protected. *Ribbed watch glasses,* when available, are more convenient to use and accomplish the same purpose.

Fig. 2.6. Rubber policeman attached to stirring rod.

The required volumetric glassware are *burets, pipets,* and *volumetric flasks.* These are illustrated and their use described in Chapter 4.

Other Equipment

In heating operations that require an ignition at moderately high temperatures (700°-900° C. maximum), *Bunsen or Tirrill burners* are generally used. Higher ignition temperatures (up to 1000° C.) may be obtained with the *Meker or fisher burner. Blast lamps* will give even higher temperatures. An electric *muffle furnace* gives the best temperature range and control. Continuous operating temperatures may be held up to about 1000° C. Their small capacity for holding crucibles, however, limits their use for large groups. Automatic *drying ovens* are used for drying samples and certain precipitates at some fixed temperature. The useful operating range is about 90°-120° C.

Platinum ware, especially crucibles, is often used in certain analytical operations. The high cost limits its use. When available, platinum ware should never be used before carefully reading the rules for its care.

Reagents

The most frequently used chemicals are of C.P. reagent quality and the commonly used dilute solutions prepared from them are as follows :

The concentration values are approximations.

Concentrated Hydrochloric Acid, 12 *N* (sp. gr. 1.18-1.19).

Dilute Hydrochloric Acid, 6 *N*. Prepared by diluting the one volume of the concentrated acid with an equal volume of distilled water.

Concentrated Nitric Acid, 16 *N* (sp. gr. about 1.42).

Dilute Sulphuric Acid, 6 N. Prepared by diluting about 380 ml. of the concentrated acid to 1 litre with water.

Concentrated Sulphuric Acid, 36 *N* (sp. gr. about 1.84).

Dilute Sulphuric Acid, 6 *N*. Prepared by pouring 1 volume of the concentrated acid slowly, with stirring, into 5 volumes of water.

Concentrated Ammonia Solution, 15 N (Sp. gr. about 0.90).

Dilute Ammonia Solution, 6 *N*. Best prepared in small quantities immediately before use unless special containers with unattackable linings are available. Preparation : 400 ml. of the concentrated reagent diluted to 1 litre with water.

The preparation of other solutions is given under the procedures for the various determinations.

For all work, distilled water should be available. Every solution that is prepared should be made uniform by stirring, shaking, or swirling it. Solutions of salts and of ammonia should be examined for insoluble matter derived from the container, stopper, or other sources. If present, the material should be filtered off.

Solutions of solids of approximate concentrations are prepared from the proper amount of material weighed on a type of sideshelf balance provided for this purpose. The pan of the balance should be protected by a watch glass, paper, or other container, depending upon the nature of the material that is to be weighted. The weighed material is dissolved first in a small amount of water and then diluted to the proper volume.

Unknowns for Analysis. The samples provided in an elementary course in quantitative analysis may be analyzed directly, without preliminary treatment. They are assumed to be homogeneous, and soluble in common reagents. The preparation, sampling, and treatment of materials not available in this form are beyond the scope of this book.

GENERAL OPERATIONS

Cleaning of Glassware and Porcelain Ware

The method of cleaning should be adapted to the substances that are to be removed. Water-soluble substances are simply washed out with hot or cold water, and the vessel is finally rinsed out with successive small amounts of distilled water. The rules for the removal of other materials should be considered for each particular case, for example, ferric oxide, manganese dioxide, and oxides in general are more easily removed by warm concentrated hydrochloric acid than by other solvents. A persistent greasy layer or spot may be removed by acetone or by allowing a warm solution of sodium hydroxide, about 1 g. per 50 ml. of water, to stand in the vessel for 10-15 minutes ; after rinsing with water, dilute hydrochloric acid, and water again, the vessel is usually clean. If not, the foregoing treatment may be repeated.

In general, glassware may be kept in a bright, clean condition by frequent washing with "Calgonite", a commercial product containing sodium hexametaphosphate. The washing solution is prepared by adding a full table spoon of the solid product to about 4 liters of water. The use of a brush aids the washing operation. Many of the detergents, commonly found in any local market, are also effective cleaning agents.

Chromic-sulfuric acid mixture, the so-called "cleaning solution", should be used when glassware, especially burets and pipets, fails to respond to the above cleaning operation. The mixture is prepared by adding concentrated sulfuric acid to sodium dichromate and allowing it to stand for some time before use. Glassware to be cleaned is brought in contact with the mixture and allowed to stand for a period of 10-15 minutes. When hot (80°-100° C.), the cleaning action is more effective. After standing, the mixture is thoroughly drained off and the glassware repeatedly washed with water. "Cleaning solution", after use, should never be thrown into the sink but should be returned to the storage receptacle. It may be used many times before it becomes ineffective. When the dark color of the mixture turns green, it should be discarded. The general use of "cleaning solution" is not recommended because of the destructive action of this mixture upon clothing, books, and laboratory furniture.

Quantitative Transfer of Liquids and Solids

One of the most frequently used laboratory operations in quantitative analysis is the transfer of substances from one vessel to another without

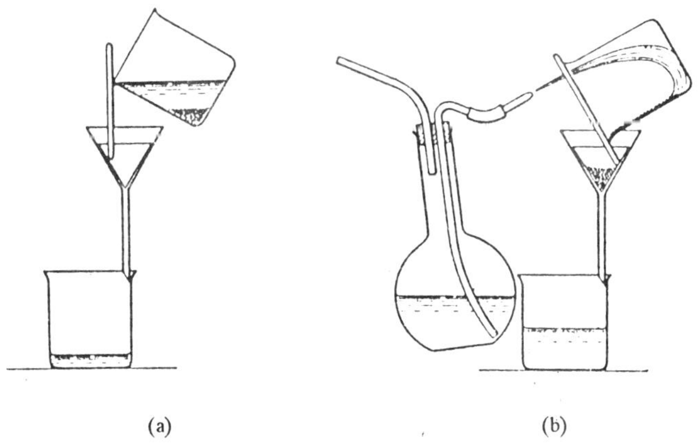

(a) (b)

Fig. 2.7. Quantitative transfer of a liquid and a precipitate. (a) Decantation of liquid through a funnel and filter paper. (b) Transfer of the precipitate to the filter.

loss. When a liquid is to be transferred from a beaker to another vessel, a stirring rod must be used in the pouring operation. See Fig. 2.7a. This prevents the loss of liquid that may run down the outside of the beaker. When the last portion has been removed, the beaker is pulled away from the stirring rod and the end of the rod touched to the wall of the receiving vessel before it is removed. Liquid remaining in the beaker is removed by repeated rinsing and transfer of the rinse liquid.

In the transfer of precipitates from a beaker to a filter, the stirring rod is pressed against the pouring lip of the beaker by the forefinger of the hand holding the beaker, while the other hand directs a stream from the wash bottle against the precipitate in the beaker. See Fig. 2.7b. The precipitate is washed down the stirring rod onto the filter. Any precipitate that adheres to the beaker is removed by rubbing it loose with the rubber policeman, after which it is washed onto the filter.

In the transfer of dry solids from a weighing bottle, the bottle is held between the thumb and fingers of one hand and a portion of the solid is removed by gentle rotation of the inclined bottle. If the fingers and thumb are clean and thoroughly dry there is no perceptible gain in the weight of the bottle when handled this way. Solids may also be transferred by means of a metal spatula. Any particles adhering to the spatula are removed by the aid of a small camel's-hair brush. Solids may be transferred from a watch glass by tipping it over a beaker of suitable size. Particles adhering to the glass are removed by a stream from the wash bottle.

Identification of Samples

In quantitative determinations, a given sample must be carried through a series of operations. In order that it may not be confused with other samples, it must be readily identified as it is carried from one operation to the next.

The original sample is generally stored in a weighing bottle. Identification numbers are usually etched on the weighing bottle so that a sample may be associated with a given bottle by the notebook record. Pencil notations on the ground areas of the bottle may also serve for identification purposes.

Flasks and beakers may be identified by Pencil notations on the ground areas available for this purpose. Special glass-marking pencils, which come in different colors, may be used on any part of the glass surfaces. Such markings, however, are not as permanent as those made with Pencil on the special areas. The use of paper labels is not recommended.

As previously stated, porcelain ware may be identified by a series of nicks made on the rim with a file, or by use of a special marking ink which permanently stains the glaze when heated. Porcelain crucibles may also be identified by a Pencil notation on the outer unglazed surface of the bottom.

Filtration with a Funnel and Filter Paper

"Ashless" filter paper is required for all quantitative filtrations where the precipitate is to be determined by ignition. This is a paper of special grade, especially treated with hydrofluoric and hydrochloric acids to remove nonvolatile matter. The paper is separated from the precipitate by burning it off a low temperature. The average ash content of a 9 cm. disc is less than one-tenth of a milligram. A correction may be applied for this, but in student work it is generally disregarded.

The speed and retentiveness of a filter paper are determined by its texture. A variety of types—medium, soft, hard, dense, thin—are available to meet the filtration requirements for precipitates of different physical characteristics.

Filter-paper pulp, when added to a solution, often aids in the filtration of fine-grained or gelatinous precipitates. The pulp is prepared by shaking with water a portion of a tablet available for this purpose. Pulp may also be prepared from quantitative paper if the tablets are not available. Here, the filter paper is torn into small pieces and boiled in acidified water until disintegration results. The use of pulp largely avoids the necessity for several grades of paper, at least as far as introductory determinations are concerned.

Glass funnels are of the type previously described. Before use, the funnel must be thoroughly cleaned to free it from all grease. The filter paper is first folded in a half-segment and then in a quarter-segment. Sometimes a small corner is torn from the outside fold of the paper, as shown in Fig. 2.8, to give a more satisfactory fit to the funnel. The paper is opened, fitted into the dry funnel, and pressed firmly into place after wetting it with water. If the operation is carried out properly, water should flow freely through the paper. There should be no air pockets formed as the stream of liquid flows down the stem. If the flow seems sluggish, a new filter paper should be fitted to the

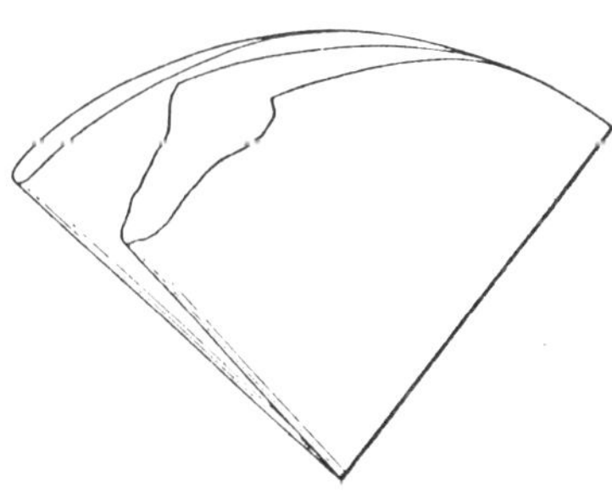

Fig. 2.8. Folded filter paper. The corner is torn off to make better contact with the funnel, behind the complete cone of the untorn portion.

funnel or another funnel should be tried. The time spent in preparing a satisfactory filtration assembly will be well repaid in time saved when it is used in any subsequent filtering operation. A slow filtration may be a lengthy and exasperating experience.

The tip of the funnel stem should touch the inside wall of the receiving vessel about 2 cm. below the rim. The filtrate is protected from possible contamination by means of a watch glass as in Fig. 2.9. In the filtration process the supernatant liquid is first decanted through the filter, the precipitate washed in the beaker and then transferred to the filter and washed again. This procedure has been previously described and will again be given in the chapter on Gravimetric Determinations.

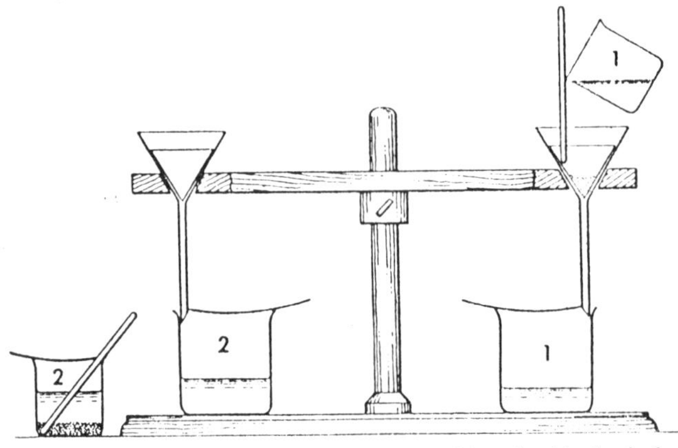

Fig. 2.9. Filtration assembly. The correct relationship of funnel and beaker is shown. The watch glass protects the filtrate from possible contamination. Note the identification numbers for the two samples.

Filtration by Suction

Under the most satisfactory conditions, filtration by a glass funnel and filter paper is usually a slow process. In certain operation filtration may be speeded up by the use of suction. Here the funnel may be fitted to a filter flask and the paper supported in the funnel by a cone of hardened paper, perforated platinum foil, or other suitable material. This procedure is not generally used in elementary courses.

Filtration by suction is most commonly carried out with some type of filtering crucible (Fig. 2.3) which is fitted to a special holder (Fig. 2.4). The completed assembly is shown in Fig. 2.10. The preparation of the commonly used Gooch crucible is given as follows.

Clean the crucibles and mark them for identification is described on page 8. Place the crucible in a suitable holder fitted to a filter flask and without applying suction pour in an amount of well shaken asbestos suspension which will eventually from a mat 0.5 to 2 mm. in thickness. The asbestos suspension is prepared by putting a special grade of the dry, acid washed, long-fibered material is a bottle until it is about one-fourth full and adding water until nearly full. The mixture is then shaken vigorously until a soup-like suspension results. The proper amount of asbestos to be added to the crucible must be found by experiment. A thin mat will filter better than a thick one and is to be preferred unless the precipitate is unusually fine-grained. If the suspension is not too thick, the crucible should be filled about two-thirds full before suction is applied.

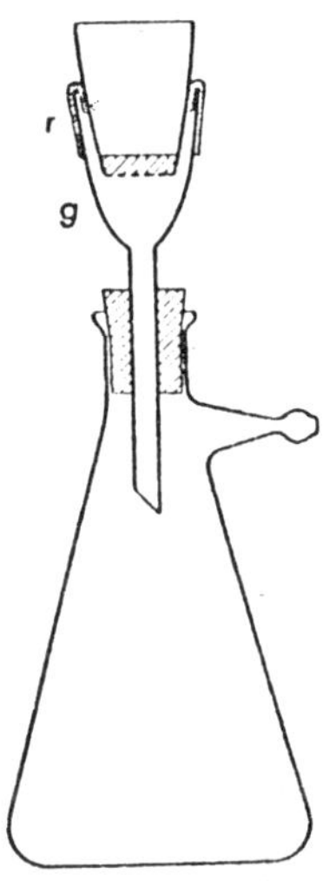

Fig. 2-10. Filtering assembly. (g) Glass filter tube holder, (r) Rubber tube so adjusted that the liquid passing through the crucible does not come in contact with the rubber.

Allow the suspension to settle for two or three minutes so that the larger particles will settle to the bottom. Apply suction gently to form an even layer of asbestos on the bottom of the crucible. The thickness of the mat may be tested by disconnecting the suction tubing from the flask while the suction is still on, removing the crucible from the holder, and looking through it in a strong light. There should be just sufficient asbestos to make the holes faintly visible. If satisfactory, replace the crucible in the holder, connect the flask to the filter pump, and add about 100 ml. of water in small portions to wash out the fine particles and soluble matter. Never pour any liquid into the crucible unless suction is being applied or the mat will be torn and run through. The liquid should be poured down a rod and allowed to fall gently on the center of the mat.

After first disconnecting the suction tubing, remove the crucible without turning off the filter pump, as described above. Wipe the outside of the crucible and dry it at the temperature that will be used subsequently in drying the precipitate. If a high temperature is to be used, most of the water is removed by careful drying at a lower temperature before the maximum temperature is applied.

Washing

In analytical operations, a precipitate must be separated from the medium in which the precipitate is formed. In some cases a precipitate free from all contamination is desired and in other the medium itself, free from the precipitate. In either instance the separation must be quantitative.

This is accomplished by washing a precipitate on a filter free from all soluble substance.

Water is generally used wherever its use is permissible, and near the boiling temperature if the nature and solubility of the precipitate will permit. Frequently it is necessary to use a special wash liquid either to make the precipitate less soluble or to prevent its conversion to a colloidal suspension. In such cases, an organic solvent is sometimes used, or more frequently, a suitable electrolyte is added to the water. The wash liquid is always added in small portions, each of which is allowed to run through the filter before the next is applied.

A qualitative test for some ion is made during the progress of the washing to check the completeness of the operation. If the filtrate is to be discarded, any convenient test may be used. If the filtrate is to be used further, the test must be one which will not add undesired substances. Experience has shown that after 6 or 8 washings the purification of the precipitate is usually complete.

Heating of Crucibles

In general, the crucible or other vessel that is to be used in the final weighing is heated initially to constant weight, at the same temperature and in the same manner that will be used later. Crucibles, other than platinum, weigh roughly 1 mg. less if brought to constant weight by heating at 700-800° C. Than if brought to apparently constant weight at 100-110° C.

Whenever crucibles are heated in a flame they should always be supported on a triangle so that the crucible is in contact with clay, silica, nichrome, or platinum. A wire gauze should never be used as a substitute for a triangle.

A ring stand that is adjustable is more suitable than a tripod. The ring, triangle, and crucible are so placed that the bottom of the crucible is just in contact with the blue oxidizing flame of the burner. The use of a luminous flame or the placing of the crucible in the cone of unburned gases should be avoided ; the soot which would form is difficult to burn off. At first, a low temperature is used to remove moisture. The flame is then increased to just char the filter paper. The carbonaceous matter which results is burned off by increasing the flame still more. In many cases, this completes the heating operation while in others the precipitate must be ignited to still higher temperatures to convert it to a weighable form. For example, a precipitate of calcium carbonate is converted at high temperature to calcium oxide.

After the heating operation, the crucible is allowed to cool somewhat and is then placed in the desiccator where it is allowed to come to room temperature before weighing. Contact of the hot crucible or its lid with the desk top, a ware gauze, or asbestos board should avoided or its lid with the desk top, a wire gauze, or asbestos board should be avoided as this may introduce an error in the weight and may leave stains that are difficult to remove.

The proper position oı a crucible and its lid is indicated in Fig. 2.11 ; with the crucible slanted in this fashion, the lid deflects a mild stream of air over the contents, thus maintaining the oxidizing conditions that are important in the ignition process.

If a blast lamp is to be used, a well-regulated oxidizing flame is applied, after a preliminary warning of the crucible. The blast flame is directed up, at an angle of about 50° to the desk top, to strike only the bottom of the crucible which is slanted up at approximately a right angle away from the line of the flame, so that the top and lid are protected from any direct draft which might cause mechanical loss of the precipitate.

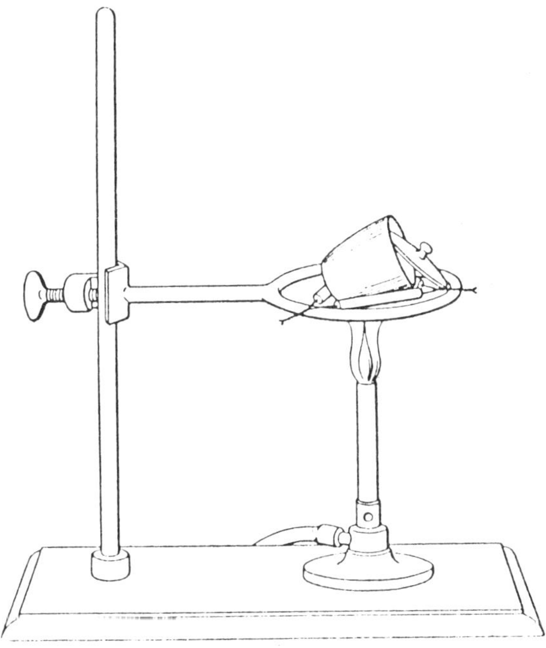

Fig. 2.11. Ignition of precipitate. This position is used after the major portion of the filter paper has been slowly carbonized with the crucible in the upright position with the lid almost covering it.

The use of a burner of the Meker type gives a high temperature without danger of mechanical loss and without attention, whereas the blast flame often varies due to pressure changes in the air line.

Electric muffle furnaces, if available, are clean and effective pieces of apparatus for igniting precipitates. Such furnaces are especially advantageous because they maintain an oxidizing atmosphere and because their temperatures can be controlled and measured.

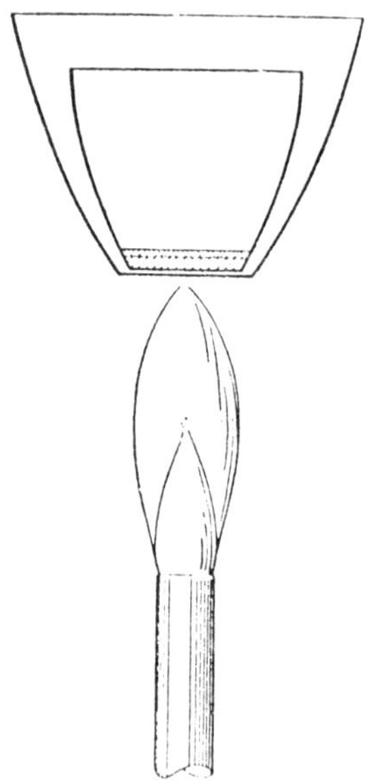

Fig. 2.12. Heating a porcelain filtering crucible. The asbestos layer and the precipitate are protected from the direct action of burner gases by the outer crucible.

As has been stated, a filtering crucible is always heated within another vessel as indicated in Fig. 2.12, unless it is heated in an electric muffle furnace. If a flame were allowed to come in direct contact with such a crucible it might cause undesirable chemical changes, such as reduction of the precipitate, and would probably injure the filtering medium.

CHAPTER 3

Measurement By Weight

THE ANALYTICAL BALANCE

Weighing in quantitative analysis is a process of comparing masses and not one of determining the attraction of gravity on a given mass. Objects of unknown mass are balanced against objects of known mass called analytical weights. The instrument for carrying out this operation is called an analytical balance and a common type used in student work is shown in Fig. 3.1.

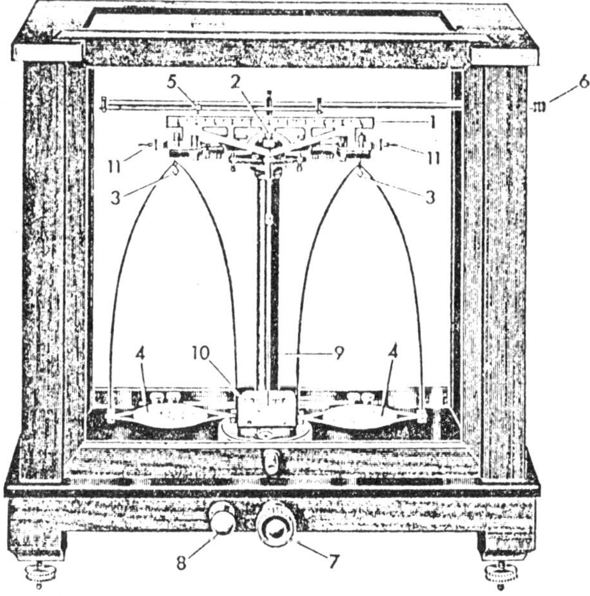

Fig.3.1. Analytical balance. 1. Graduated beam. 2. Center knife-edge. 3. Supporting stirrups for pans. 4. Pans. 5. Rider hook. 6. Rider rod. 7. Beam-release knob. 8. Push-button pan arrest. 9. Pointer. 10. Pointer scale. 11. Adjusting screws for pointer.

The working part of the balance is enclosed in a glass-fitted case which is made either of wood or an aluminum alloy. The baseplate is usually of black glass or black slate. The front of the case is sliding door which lifts upward and is held in position by weights concealed in each side of the case.

The beam (1) is a lever of the first class with a knife-edge fulcrum (2) made of agate which rests on a plate (bearing) also made of agate. An enlarged section is shown in Fig. 3.2. The fulcrum is located as nearly as possible equidistant from the terminal knife-edges which do not show clearly in the diagram and are not indicated. Near each end of the beam are hung stirrups (3) which support the bows on which are fastened the

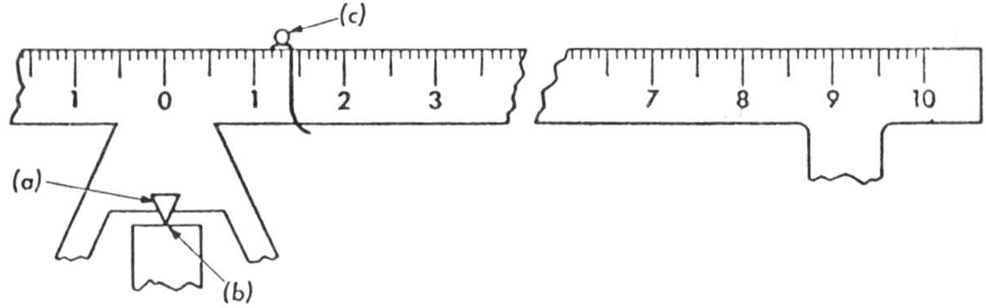

Fig. 3.2. Section of beam. (a) shows center knife-edge, and (b) the agate bearing plate on which it rests. (c) the rider shows a reading of 1.4 mg.

pans (4). The beam is usually graduated into 10 numbered divisions on each side of the fulcrum. Each numbered division is further subdivided into 10 divisions so that there is a total of 100 divisions on each side of the beam. Each small division corresponds to a weight of one-tenth milligram when the rider is placed on some position on the scale. The rider is a piece of aluminum wire with a loop at the top and bent in the bent in the form of an inverted U, as in Fig. 3.4b. A rider hook (5), attached to the rider rod (6), permits the rider to be placed on any scale division.

On the lower front of the platform on which the case rests is a milled knob (7). When this knob is turned it lifts the beam off the knife-edges and the loosely hung stirrups are held firmly in place. A push-button knob (8) to the left of the beam-release knob, controls the pan arrests. These are arms concealed under each pan. With the push-button knob in, the pan arrests lightly support the pans to prevent them from moving when the beam is lowered on the knife-edges. These pan arrests usually yield when the load on one pan is a gram or more heavier than on the other one.

If the beam is lowered on the knife-edges and the pan arrests released, the beam and the pans will move in the manner of a compound pendulum— that is, a pendulum with harmonic damped oscillation. The movement is observed by means of a pointer (9) which moves across the pointer scale (10). The scale is white in color and on it are ruled usually

20 divisions, each one a millimeter in width, with 10 on each side of the center line. With perfect adjustment the pointer will swing an equal distance on each side of the center line. Such a condition seldom occurs nor is it necessary. Adjustment may be made with the two adjusting screws (11), but should never be done without consulting the instructor.

The Rest Point

With the balance free to the swing, the pointer will eventually come to rest at position on the scale. This position is known as the rest point of the balance. It may be on the center line or a position to the right of left of the center line. Since waiting for the balance to reach the rest point is time-consuming (unless the balance has a damping device), the rest point may be calculated from a few successive readings of the turning points of the pointer.

The procedure is as follows: Lower the beam on the knife-edges, release the pan arrests and start the pointer swinging. This usually occurs on release of the pan arrests if the adjustment is off dead center on the scale. Should the pointer fail to move, motion may be started by wafting air gently against one of the pans by a movement of the hand. As the pointer moves across the scale, its furthest position on each side is read by means of arbitrarily-assigned numbers as indicated in Fig. 3.3. An average reading of successive swings is calculated for the right and left turning points since an estimated position of the moving pointer for a single swing is subject to uncertainty. The rest point is the median of the two average readings. Usually the first two swings after releasing the beam are disregarded. An odd number of turning points is read to give a whole number of complete swings.

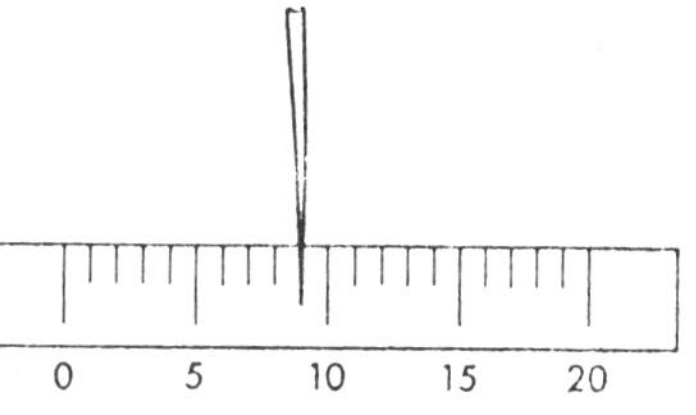

Fig. 3-3 Pointer scale and pointer.

Example:

	Left Readings		*Right Readings*
	4.3		14.0
	4.7		13.7
	5.0		
Sum..............	14.0	Sum..................	27.7
Average...........	4.7	Average...........	13.9

$$\text{Rest Point} = \frac{4.7 + 13.9}{2} = 9.3$$

The rest point of an empty balance may vary frequently and should be determined before each series of weighings. Sometimes the scale is numbered differently: the center line, numbered 10 on the diagram, is taken as zero; the line numbered 0 is given the value –10 and the line numbered 20 the value of +10. The method suggested here is usually preferable for beginners since there is less chance for error than when – and + numbers are used.

Sensitivity

If the rider is placed on the one-milligram division of the beam and the rest point now determined with this load, the rest point will be different from that determined on the empty balance. The difference between the scale readings for the two rest points may be defined as the sensitivity of the balance.

Example:

Rest point of the empty balance.............................. 9.3

Rest point with rider at 1 mg................................... 6.3

Displacement of rest point............ 3.0 divisions per mg.

The sensitivity of a balance varies with load. For example, if the sensitivity is determined with a load of say 10 grams on each pan, the value obtained will be slightly less than that obtained with the empty balance. As will be explained later, the sensitivity is often used in a weighing operation. For this reason, the sensitivity may be obtained for a series of different loads, say 10 g., 20 g., 30 g., 40 g., and the values plotted on a chart as sensitivity *vs.* load. Such a chart may be kept as a permanent reference record in the laboratory notebook since the sensitivity of a given balance, once determined, will not change unless the balance is damaged.

The manufacturers of balances usually define sensitivity as the smallest change in load that a balance will detect with maximum load, which is 200 grams on each pan. To say that the sensitivity is 1/10 milligram means that an excess of one milligram on one pan with maximum load will result in a change in rest point of 2.5 divisions. Since the uncertainty in reading the scale is 0.25 divisions, 1/10 milligram is the smallest change in weight that can be detected with certainty. The sensitivity of analytical balances are usually rated as 1/10 mg., 1/20 mg. or 1/40 mg. Only microanalytical balances exceed this range of sensitivity.

Weights

A box of typical analytical weights is shown in Fig. 3.4a. There are nine gram weights: 100 g., 50 g., 20 g., 10 g., 10 g., 5 g., 2 g., 2 g., and 1 g. Sometimes the 100 gram weight is omitted. The less expensive weights are made of lacquered brass while those of better quality are either made of a type of stainless steel or are rhodium plated. There are also eight or more fractional weights. There are: 500 mg., 200 mg., 200 mg., 100 mg., 50 mg., 20 mg., 20 mg., 10 mg. Sometimes 5 mg., 2 mg., and 1 mg. weights are included but these are not necessary if values of 10 mg. or less can be obtained on the beam scale with the rider. A modern trend is to include 3 mg., 30 mg., and 300 mg. in the fractional weights and 3 g. and 30 g. in

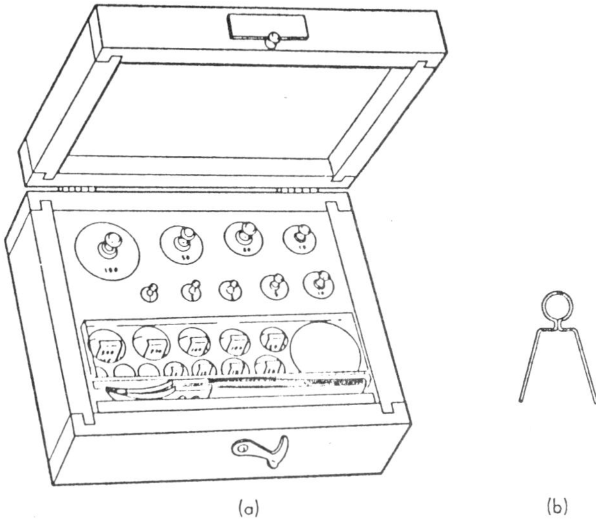

Fig. 3.4. (a) Box of analytical weights. (b) Balance rider.

the gram weights so that no duplication of any weight value is necessary. All of these fractional weights are commonly made of aluminium. The 500 mg. to 50 mg. weights are also made of tantalum. All weights should be handled only with the ivory or plastic-tipped forceps provided. Weights of 10 mg. to 0.1 mg. are obtained from the beam by the proper position of the rider.

Some balances may be graduated with only 50 divisions on each side of the center of the beam. It is important to note that here a 5 mg. rider must be used rather than a 10 mg. rider. When a rider is to be replaced on a balance it is always advisable to compare it with a known weight to be sure that it is the proper one to be used with the balance.

Weights come in different degrees of accuracy. Two common National Bureau of Standards specifications are Class S and Class S-2. Tolerances for the latter may be as much as five times that for the former. By tolerance is meant the allowable possible deviation from the stamped value on the weight. For example, in Class S weights the tolerance for 2 and 1 gram weights is 0.10 mg. It increase for the other gram weights and decreases for the fractionals. For student and routine analytical work, Class S-2 weights are generally used.

When any uncertainty exists in the accuracy of a set of weights, or if they are to be used in work of the highest accuracy, they should be calibrated.

Where a set of weights may be commonly used by a number of individuals there is little purpose in calibrating them since weights from one set may be carelessly transposed to another.

DIRECTIONS FOR WEIGHING

First determine the rest point of the balance. Raise the beam off the knife-edges and place the object to be weighed on the left pan. With the forceps, select from the weight box a weight that is thought to be slightly heavier than the object and place it on the other pan. Carefully lower the beam on the knife-edges. If the weight added is much heavier than the object, the pointer will start to move in the direction of the object; if much too light, the pointer will move in the other direction. Should the pointer fail to move, release the pan arrests gently to observe the direction of movement. This will become necessary as the mass of the object and that of the weights approach each other. If the weight is too heavy, return it to the weight box and replace it with the weight of the next value; if too light, leave it on the pan and add a weight of the next value. Each time a weight is added or removed, the beam must be raised off the knife-edges by the beam release knob, otherwise the loosely hung stirrups on which the pans are hung may be jarred out of adjustment.

This systematic procedure is used for each successive weight in the weight box. Do not skip from one weight to another. When the weights from the weight box have been exhausted, close the balance case, lower the beam on the knife-edges and leave it there until the weighing is completed. Place the rider on the 5 mg. division of the beam, release the pan arrests, and observe the movement of the pointer. If the movement is directly toward the pan holding the object the weight is too heavy, and if in the other direction, it is too light. Next, try the position halfway between O mg. and 5 mg. or halfway between 5 mg. and 10 mg. Continue this systematic procedure until the rest point obtained is the same as that originally determined with the empty balance. Add up the weights on the pan and to this sum add the weight by the position of the rider. Record the value in the notebook. Return the weights to the weight box, but again check each one to be sure that no error was made in the previous summation. Check the position of the rider, lift it from the beam, and close the balance case.

While this trial and error procedure is laborious and time-consuming, it may be preferred as the introduction of a beginner to the use of an analytical balance.

The use of the sensitivity of the balance in the weighing process is a more rapid and often preferred procedure. In this method, follow the same procedure as described above until the approximate weight is located so that the pointer will come to rest within 10 divisions on either side of the center position on the scale. Determine this rest point accurately. The difference between this rest point and that determined for the empty balance gives the number of divisions that the pointer is displaced. Using the sensitivity for this load, calculate the weight that must be added or subtracted to bring the pointer to the rest point for the empty balance.

Example:

Weight on right pan....................................	10.1580 g.
Rest point for this load................................	12.0
Rest point for empty balance.......................	10.5
Displacement of rest point...........................	1.5 divisions
Sensitivity for 10 gram load.........................	3.6

$$1.5 \div 3.6 = 0.4 \text{ mg.}$$

Since the rest point of the loaded balance is displaced to the right of the center scale division, this weight must be *added* to the weight on the pan to give the correct weight of the object: 10.1580 g. + 0.0004 g. = 10.1584 g. Check against a possible error in computation by placing the rider on the calculated position on the scale (10.5 in the above example).

As one gains in experience, weighing becomes a less time-consuming operation. In routine weighing operations, an experienced analyst often allows the pointer to make a single short swing of about 6 to 8 divisions and mentally estimates the midpoint of the swing. With a short swing, the frictional effect is less than that for a longer swing so that even if more than one successive short swing were made, differences in the estimation of the midpoint would be small enough to be disregarded. The best accuracy, however, is always obtained by the calculation of the rest point in the manner described.

Other Types of Balances

In student work the rider type of balance is in common use because it is less expensive and because it has a simpler mechanism than other types. Most analysts prefer other types of balances, since weighing with these is more rapid. The principal difference between the rider type and other types is in the manner in which the fractional weights are added. The chainomatic balance uses a chain, one end of which is attached to the beam and the other to a sliding vernier scale. See Fig. 3.5. A handwheel on the lower right of the balance case moves the vernier along an upright post placed in front of the right-hand pan. The post is ruled into 100 divisions each of which corresponds to one milligram. The vernier gives the position between each division so that the total load that may be measured by this device is from 100 mg. to 0.1 mg. Sometimes the post is mounted in a horizontal position at the top front of the balance, while in other the post is repealed by a large dial mounted above the right-hand pan.

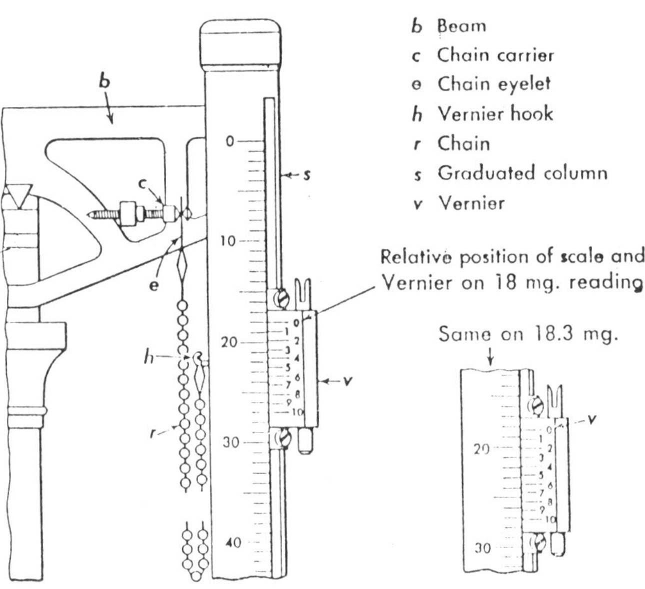

Fig. 3.5. Graduated column and vernier of chainomatic balance. (Courtesy of Christian Becker, Inc.)

The position of the chain is determined by rotating the dial by a handwheel. Other types of balances have a keyboard arrangement which automatically adds or removes weights by depressing keys located on the lower front of the balance case. The weights may be transferred back and forth while the beam is in motion. A more complete description of the many modifications of analytical balances is given in the manufacturer's catalogs.

Balance Accessories

A magnetic damping device is supplied with many balances or may be installed in any balance. The device, which is attached to a suitable part of the balance, is a metal which is suspended between the poles of a permanent magnet. The purpose of the damping device is to slow down the oscillations of the pointer when it is free to swing, so that the rest point may be determined more quickly.

Portable magnifiers that may be placed in front of the scale assist in reading the movement of the pointer by reducing eye strain.

Balance desiccators are small containers that hold an appropriate desiccant such as anhydrous calcium sulfate (Drierite), or silica gel. They are placed inside the balance case to maintain a dry atmosphere.

Counterpoised watch glasses may be used as a protection for the pans and for the direct weighing of powdery or crystalline solids. They are supplied in numbered pairs, each watch glass of equal weight.

Practice in Weighing

To obtain familiarity with the balance and the method of weighing, some preliminary practice is desirable. Among other methods, two simple exercises are suggested:

1. Various test pieces, such as numbered sheets of brass of different sizes, are weighed and the values are checked with the instructor.
2. A nonhygroscopic salt is placed in a weighing bottle; the bottle and contents are weighed. A portion of the salt is removed carefully to another weighing bottle, or to one of a pair of counterpoised watch glasses. The weight of the sample removed is determined directly and by difference. The two weighs should agree to 0.0002-0.0003 g.

RULES IN WEIGHING

1. Students assigned to a balance are responsible for the neatness of the balance and its immediate surroundings. No weighing should be done on the analytical balance that can just as well be done on a less-sensitive, side-shelf balance.
2. The balance case door is kept closed when the balance is not in use in order to protect it from dust and fumes. It is also closed in the final adjustment of weights and while swings are being observed, to avoid drafts.

3. Neither weights nor other objects shall be placed upon or removed from the balance pans unless the beam in raised off the knife-edges and the pan arrests are in place. This is the most commonly-violated rule with beginners and invariably results in the dislocation on the beam of the stirrups from which the pans are hung.

4. The beam support and pan arrests must be raised and lowered gently to avoid damage to the agate knife-edges.

5. No object is to be weighed unless it is at room temperature. Heated objects placed in the balance case cause air currents which result in an erratic behavior of the balance. Also, any heating will cause an unequal expansion of the balance mechanisms.

6. No powder or crystalline material must ever be placed directly on the balance pans. Use a watch glass, weighing bottle, or other suitable container.

7. Never overload a balance. For most analytical balances, the maximum capacity is 200 g. on each pan.

8. In any weighing operation always check the individual values and the summation of the values of the recorded weights. Errors in reading weights are easily made.

9. The weights should be handled only with the forceps provided. When the weighing operation is completed, return all weights to their proper positions in the weight box. If a particular weight is missing, consult the instructor. Never borrow a weight from another weight box. The rider should be lifted up on the rider carrier when the weighing is finished, or placed at the zero position on the beam.

10. A balance out of adjustment (observed by erratic behavior), or weights or a rider missing should be immediately reported to the instructor.

CHAPTER 4

Measurement by Volume

TYPES OF VOLUMETRIC APPARATUS

The instruments generally used in the quantitative measurement of volume are burets, pipets, and volumetric flasks. The usual units of measurement are the milliliter and the liter.

The precision of volumetric work depends upon the accuracy with which volumes of liquids can be measured. This is much less than the accuracy of weighing. There are certain sources of error which must be carefully considered. Changes in temperature cause changes in the volume of glass apparatus and liquids. In ordinary glass flask of 1000 ml. volume increases in capacity 0.025 ml. per degree centigrade, but if made of Pyrex glass, the increase is much less. An increase 10°C. At room temperature causes a volume increase of 0.20 ml. for 1000 ml. of water or most dilute, aqueous solutions. There may be errors of calibration of the apparatus, that is, the volume marked on the apparatus may not be the true volume. Such errors can be eliminated only by recalibrating the apparatus.

Burets (Fig. 4.1) are used to deliver variable volumes. They are long tubes of clear glass provided with a stopcock at one end and generally graduated in units of one-tenth millilitre. The graduations usually cover a range of fifty milliliters. For the better burets, the graduations encircle the buret at intervals of one milliliter.

Volumes are measured by filling the buret with liquid to the upper part of the graduations, reading the level of liquid, and then reading the level again after the liquid has run out through the stopcock. Use of a buret reader is desirable to determine the position of the liquid level in relation to the graduations. A common type of reader is in the form of a card, made of opaque plastic material, that may be attached to the buret. The upper half of the card is white and the lower, black. In reading the buret, the dividing line is adjusted so that it coincides with the meniscus of the liquid when viewed at the same level as in Fig. 4.2. Parallax is avoided by adjusting the position of the buret to the eyes so that the front and rear of the encircling line nearest the meniscus appear to coincide. For dark-colored liquids, such as potassium permanganate, the top of the level of

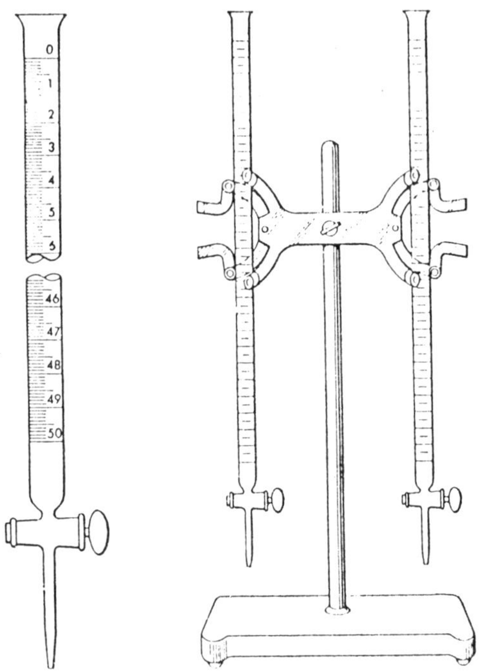

Fig. 4-1. A buret. Fisher holder with burets.

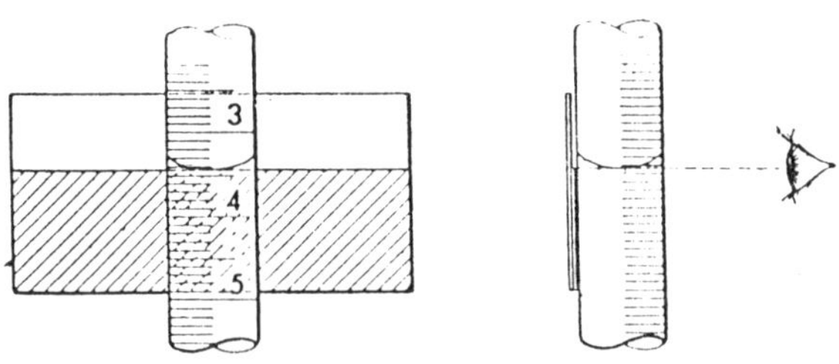

Fig. 4-2. Illustrating the use of buret reader.

liquid is read. Since volumes are always measured by a difference in reading, exactly the same procedure should be used in making the two readings.

The stopcock must be properly lubricated with special stopcock lubricant to permit case of rotation and prevent leakage. To lubricate a stopcock, remove the stopper from the buret, wipe dry and apply a very small amount of lubricant as shown in Fig. 4.3. Insert the stopper in the buret, press firmly, and rotate to get a continuous film of the lubricant. A common tendency is to use too much lubricant. This excess gets in the tip and clogs the flow of liquid. It may be removed by a thin wire probe or by means of some solvent such as acetone or ether. The stopper should be attached to the buret by means of a "stopper tie" or a wire.

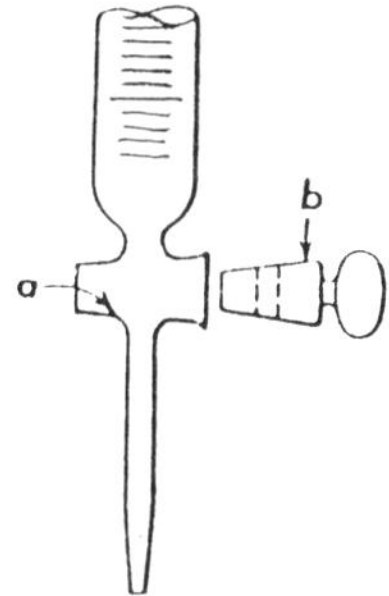

Fig. 4.3. Lubrication of stopcock. The lubricant is applied lightly at points *a* and *b*.

Before use, the buret must be perfectly clean. This is best accomplished by pouring into the buret 10-20 ml. of a solution of Calgonite, rubbing a little soap on a buret brush, inserting the brush in the buret and pulling the brush back and forth about a half-dozen times. The buret is thoroughly rinsed out with water and a test for cleanliness made by observing whether droplets of water adhere to the inner walls when water is run through it. If this treatment is ineffective, "cleaning solution" should be tried.

Do not dry a buret that as been cleaned for use, but rinse it two or three times with a little of the liquid with which it is to be filled, and discard the washings. Do not allow alkaline solutions to stand in a buret, because the glass will be attacked and the stopper will tend to stick. A buret will remain clean for a longer period of time if it is kept filled with water when not in use.

To use a buret, fasten it to a buret holder and lower it so that the tip is within the opening of the receiving vessel. This will prevent spattering of the liquid as it leaves the buret. The stopcock is manipulated by the thumb and finger of the left hand while the right and rotates the receiving flask, or a stirring rod if a beaker is used (Fig. 4.4). If properly lubricated, the stopcock should turn with slight resistance as pressure from the thumb and finger is applied. With practice the flow of liquid may be completely controlled. Care should be taken not to push out the stopper with the palm of the hand. Stoppers tend to slip out easily, causing liquid to seep out between the stopper and the buret. Frequent pressure should be applied to the stopper to be sure that it is always firmly in place. In the final removal of liquid from the buret, touch the tip to the wall of the receiving vessel and wash down the liquid with a stream from the wash bottle. It is possible to split drops by removing liquid from the tip by means of a stirring rod. The delivery time should be adjusted to approximately 0.7 per second for dilute aqueous solutions, otherwise too much liquid will adhere to the walls and the volume delivered will be too small. A reading would not remain constant, but the meniscus would gradually rise as the liquid drains down the walls.

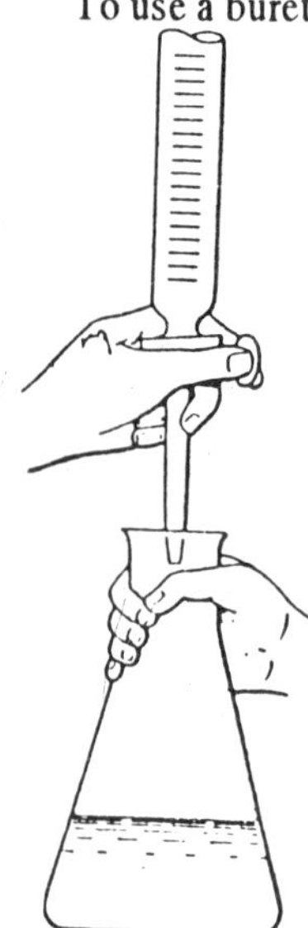

Fig. 4-4. Manipulation of buret in titration. The flask is gently swirled with one hand while operating the stopcock with the other in such a manner as to hold the plug in place to prevent leakage.

In laboratories where many titrations are to be carried out, a special type of buret is attached to the stock bottle of solution. A siphon arrangement permits the automatic filling of the buret.

Volumetric Pipets (See Fig. 4.5) are instruments designed to deliver a fixed volume. The usual capacities are 5, 10, 25, and 50 ml. Before use, they must be throughly cleaned. This is not as simple an operation as with a buret, "Cleaning," preferably hot, is recommended. The pipet is dipped in the "cleaning solution" which is sucked up so as to be above the etched mark on the stem. Care should be used in drawing up this corrosive solution as well as any alkaline or poisonous liquids. It is available to attach a rubber tube to the top of the pipet and apply suction from the month through this. The "cleaning solution" should be completely removed by several rinsings with water, and a test for cleanliness made by observing whether droplets of liquid cling to the inner walls when water is run from it. The pipet should then be rinsed several times with the liquid to be measured and the washings discarded.

Now a sample may be taken by inserting the tip in the liquid and sucking up the liquid to above the etched mark on the stem. The upper end is closed with the dry index finger, any adhering liquid is wiped from the outside of the lower stem and the level is allowed to fall slowly by regulating the pressure on the finger until the bottom of the meniscus is tangent to the mark. Liquid clinging to the tip is removed by touching the tip to the side of the container and the pipet is allowed to empty freely into the receiving vessel. When the free flow has stopped, the tip is touched against the wall of the receiving vessel and removed.

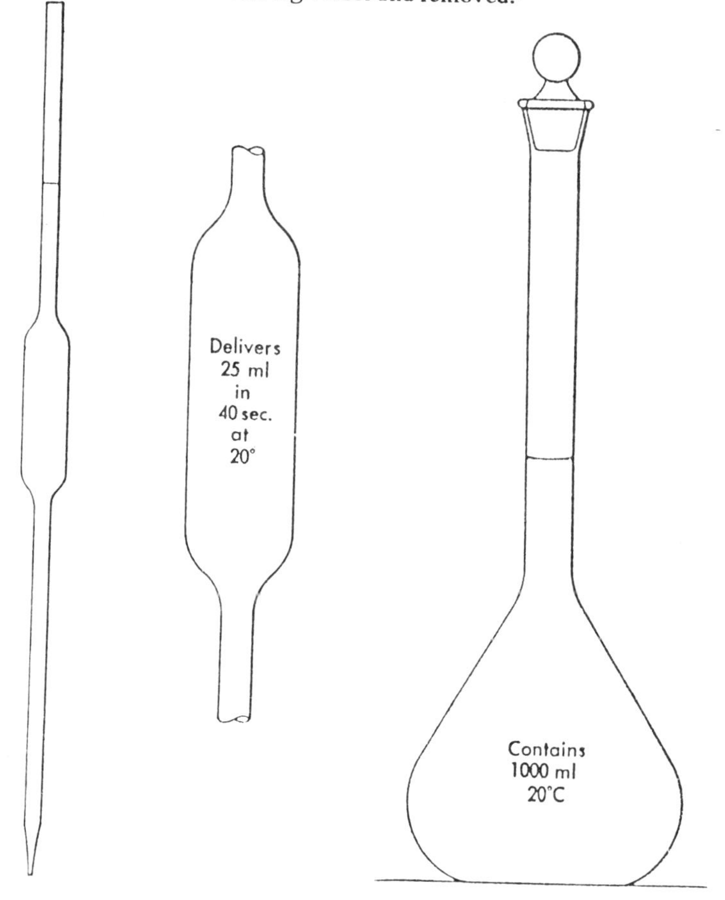

Fig. 4.5. Volumetric pipet.

Fig. 4.6. Volumetric flask.

The pipet is designed to deliver a specific volume of liquid when the above procedure is followed. A certain amount of liquid should remain in the tip. *This is never blown out* except in the case of certain types of pipets so designated.

Volumetric Flasks (See Fig. 4.6) are usually designed to contain a given volume when filled to that the meniscus is tangent to the ring which has been etched on the neck of the flask, the eye being at the level of the ring. The commonly-used capacities of these flasks are, 250, 500, and 1000 ml. A measured amount of the substance to be diluted to the specific volume is first diluted in a beaker and then transferred quantitatively to the flask. Water is added to the mark, the glass stopper inserted and, the flask shaken to ensure homogeneity of the solution.

CALIBRATION OF VOLUMETRIC APPARATUS

For the usual analytical quality volumetric apparatus, the limits of error, designated as tolerance, are given in the table below. For precision-grade apparatus, which meets the specifications of the U.S. Bureau of Standards, the tolerances are somewhat smaller.

Burets

Capacity, ml	25	50
Tolerance, ± ml	0.06	0.10

Pipets

Capacity, ml	5	10	25	50
Tolerance, ± ml	0.02	0.04	0.06	0.10

Flasks

Capacity, ml	250	500	1000
Tolerance, ± ml	0.24	0.30	0.60

When using apparatus of good quality if may not be desirable for beginners to attempt calibration of such apparatus. The reasons for this are as follows: (*a*) with limited time available, the operations are time-consuming; (*b*) with inexperience in analytical technique, the "calibrated" apparatus may be less accurate than the apparatus calibrated by the manufacturer; (*c*) the assumed tolerances are usually less than the expected error permitted in a determination by a beginner. If calibration seems desirable, the procedures are outlined below.

Calibration of a Buret

Thoroughly clean the buret and seen that the stop-cock is properly lubricated. The buret is then filled with distilled water which has come to room temperature, and the temperature noted. The water delivered for convenient intervals, say 5 or 10 ml., to the capacity of the buret, is caught in a weighing bottle and weighed. Weights need be observed only to the nearest one-hundredth of a gram. The readings should be estimated to the nearest hundredth of a milliliter. All observations should be tabulated. The volume actually delivered for each interval is obtained by multiplying the weight of this volume occupied by one gram of water at the temperature in question.

Calibration of a Pipet

Fill the thorough-cleaned pipet with distilled water at the noted temperature and allow it to deliver into a previously weighed bottle or a 50 ml. glass-stoppered flask.

Calibration of a Volumetric Flask

Flasks may be calibrated in a manner similar to that described for burets and pipets. However, their large volume requires the use of a special balance of high capacity and special weighing technique.

Approximate Comparison of Flask and Pipet

In analytical operations it is frequently desirable to prepare a single large sample, sometimes called a master sample, and then take a known fraction of this for analysis. This fraction is known as an *aliquot.* The sample is diluted in a volumetric flask and the aliquot taken with a pipet. A rough check of the volume of one vessels in terms of the other may be obtained as follows: Prepare a clean 50 ml. pipet and a clean, *dry* 250 ml. volumetric flask; fill the pipet to the mark and empty it five times in succession into the flask. if the level does not upper edge on a level with the bottom of the meniscus. The volume of the flask now bears a direct relationship to the volume of the pipet.

VOLUME MEASUREMENTS IN ANALYSIS

Volumetric analysis or analysis by titration is concerned with the relationships in the measurement of the volume of a solution required to react with a volume of a second solution or with a given weight of a material.

The process of measuring the volume of a solution that is required to complete a reaction is known as *titration.* The relationship at which some detectable evidence indicates the completion of the reaction is called the *end point.* The end point should coincide as closely as possible with the relationship at which chemically equivalent amounts of the regent and the substance titrated have been brought to a completed reaction. The latter is called the *equivalence point.*

A reaction must fulfil the following conditions to be suitable for use in analysis by titration: (1) it must go practically to completion when chemically equivalent amount of the reacting substances are present: (2) it must be practically instantaneous; (3) the end point must be sharply defined by some change in the physical or chemical properties of the solution, such as a change in colour or the formation of a precipitate. When the reagents themselves do not make this possible, a third substance, known as an indicator, is added which reacts only when the main reaction is completed.

The number of reactions suitable for analysis by titration is comparatively small and many reactions which are otherwise satisfactory cannot be used because there is no indicator which will show when the equivalence point is reached.

Titration methods are generally preferred over gravimetric methods because they are more rapid. Gravimetric methods often involve tedious and difficult separations and time-consuming evaporations and ignitions. Also most titration methods are more accurate than gravimetric methods when the latter are subject to errors that cannot be easily avoided.

The reaction methods fall into three general classes as follows: (1) *Neutralization reactions,* which are reactions involving the union of hydrogen and hydroxyl ions. These are referred to as acid and base titrations or acidimetry and alkalimetry. (2) *Oxidation-reduction reactions,* which involve a change in the oxidation state of constituent elements in the reacting substance. They are referred to as redox titrations. (3) *Precipitation reactions,* in which a precipitate is formed. No change in oxidation state occurs nor are hydrogen and hydroxyl ions involved. Also considered with these are reactions in which a complex ion is formed.

Procedures and the discussion of determinations for each of these classes are given in the appropriate chapters that follow.

SYSTEMS OF EXPRESSING SOLUTION CONCENTRATION

There are many systems of expressing the concentration of solutions. In analytical chemistry, the most commonly used are: (1) weight of solute present in a given volume of solution; (2) moles of solute present in one liter of solution, called molarity; (3) gram equivalents of solute present in one liter of solution, called normality; (4) percentage of solute by weight in a solution of known specific gravity.

Weight of Solute per Unit Volume of Solution

In the preparation of a solution using this system, a known weight of the solute is diluted to a known volume. Where accuracy is required, the solute is weighed on an analytical balance, transferred to a volumetric flask, and diluted to the known volume. The concentration is usually expressed as grams of solute per milliliter of solution. Any multiple of fractional part of such a solution will contain a known weight of solute. For example, if 40.00 grams of sodium hydroxide is dissolved in one liter of solution, then each milliliter of the solution will contain 0.0400 gram of sodium hydroxide; 25.00 ml. would contain 1.000 g.; 35.25 ml. would contain 1.410 g.

Molar solutions

Molarity is a system of expressing concentration based on the number of moles of solute present in one liter of solution. Since the liter is a large unit of volume, it is more practical to define molarity in terms of millimoles of solute present in the milliliter of solution, where one millimole is a molecular weight in grams divided by 1000.

If the concentration of a solution is known in terms of grams of a given solute present in a known volume of solution, the molarity may be calculated.

Example : If 375.0 ml. of a solution contains 5.500 g. of NaOH, what is the molarity?

Divide the weight of solute by the millimolecular weight of the solute. This gives the number of millimoles of solute present. Divide this result by the volume in which it is dissolved and the result is millimoles per milliliter or the molarity.

$$\frac{5.500}{0.04000} = 137.5 \text{ and } \frac{137.5}{375.0} = 0.3667$$

The concentration of the solution would be designated as 0.3667 M.

If the molarity of a given solution is known, then the concentration of the solution expressed as grams of solute per unit volume of solution may be calculated.

Example : How many grams of NaOH are present in 165.0 ml. of a 0.2350 *M* solution of sodium hydroxide?

Multiply the molarity by the millimolecular weight of the solute. The result is the grams of NaOH present in each milliliter. Multiply this by the given volume and the result is the weight of solute present.

$$0.2350 \times 0.04000 \times 165.0 = 1.551 \text{ gram of NaOH}$$

Normal solutions

Normality is a system of expressing concentration based on the number of equivalents of solute present in one liter of solution or the number of milliequivalents of solute present in one milliliter of solution. The only difference between this system of expressing concentration and that of molarity is that the unit of measurement of the solute is the equivalent or milliequivalent instead of the mole or the millimole.

One equivalent of a substance is either one mole of the substance or some whole-number fraction or whole-number multiple of one mole, depending on the nature of the reaction in which the substance participates. In reactions between acids and bases, one equivalent is the weight of regent which contains or reacts with one gram atom of replaceable hydrogen (1.008 g.) or with one gram molecule of hydroxyl (17.008 g.) For precipitation reactions, complex formation reactions, and oxidation-reduction reactions, the definition of an equivalent is different and will be discussed in the appropriate chapters.

For the reaction,

$$HCl + NaOH \rightarrow NaCl + H_2O$$

one equivalent of HCl or NaOH is the same as the mole. On the other hand, in the reaction,

$$H_2SO_4 + 2NaOH \rightarrow Na_2SO_4 + 2H_2O$$

one equivalent of H_2SO_4 is $\frac{H_2SO_4}{2}$, but that of NaOH is $\frac{NaOH}{1}$.

For the reaction between hydrochloric acid and sodium carbonate, one equivalent of sodium carbonate may be $\frac{Na_2CO_3}{1}$ or $\frac{NaCO_3}{2}$, depending on whether the reaction is:

$$Na_2CO_3 + HCl \rightarrow NaHCO_3 + NaCl$$

or

$$Na_2CO_2 + 2HCl \rightarrow H_2CO_3 + 2NaCl$$

In the example on the calculation of molarity, the normality of the NaOH solution would be the same as the molarity, and the concentration of the solution could be designated as 0.3667 *N*. However, if $Ca(OH)_2$ had been the base used in the example, then the normality of the solution would be different than the molarity.

Example : What is the normality of a solution of calcium hydroxide that has a concentration of 6.250 g. of $Ca(OH)_2$ in 225.0 ml. of solution ? Divide 6.250 by 0.03705 (the milliequivalent weight of calcium hydroxide). The result, 168.6, gives the number of milliequivalents present. Divide this by 225.0 and the result, 0.7493, is the number of milliequivalents of $Ca(OH)_2$ per milliliter, or the normality.

Note that the molarity of this solution would be 0.3746, half the value of the normality.

Concentrations of solutions expressed in terms of normality are commonly used in analytical work involving titrations. The general practice of designating solutions in terms of normality, when not used in titrations, is not a good one. The reason for this is, as previously stated, that the equivalent weight of a given solute varies with the reaction in which it participates. For use other than titrations, it is better that solutions be designated in terms of molarity . Much confusion may thus be avoided.

If the volumes required for reaction of two solutions are known and the normality of one of these is known then the normality of the other may be calculated by the simple relationship:

Millilitres × normality (solution No. 1)

= millilitres × normality (solution No. 2)

Example : If 25.20 ml. of 0.702 *N* acid reacts with 35.10 ml. of base, what is the normality of the base?

$$25.20 \text{ ml.} \times 0.2703\ N \text{ (acid)} = 35.10 \text{ ml.} \times N \text{ (base)}$$

$$N \text{ (base)} = 0.1941$$

To find the number of milliequivalents present in a given amount of any sample, apply the following rules.

Rule 1. Divide the weight in grams of a pure substance by the milliequivalent weight in grams of the substance, where the milliequivalent weight is determined by the nature of the reaction involving the substance.

Rule 2. If the sample is a given volume of a solution of known concentration, calculate the normality of the solution (milliequivalents of solute per milliliter of solution) and multiply this by the given volume.

Concentration Expressed as Specific Gravity and Percentage by Weight.

The concentrations of certain common laboratory regents are frequently expressed in terms of specific gravity and percentage of solute by weight. These reagents include:

1. Concentrated hydrochloric acid, sp. 1.18–19, 35–38% HCl
2. Concentrated nitric acid, sp. gr. 1.42, 70% HNO_3.
3. Concentrated sulfuric acid, sp. gr. 184, 95.5% H_2SO_4.
4. Concentrated ammonia solution, sp. gr. 0.90, 27-30% NH_3.

Tables in handbooks give the corresponding values for these as well as certain other solutions in a wide range of dilutions.

For most purposes, specific gravity may be defined as density (weight in grams of one milliliter) and is used in this sense in all calculations given here. If the specific gravity of a solution is multiplied by the percentage of solute present, the result gives the concentration in terms of grams of solute present in one milliliter of solution.

Example : 1.84 × 0.955 gives an approximate concentration of 1.76 grams of H_2SO_4 per milliliter for concentrated sulfuric acid. Expressed as molarity, the concentration would be 18 *M*, and as normality, 36 *N*.

CALCULATIONS

While the basic calculations in volume measurements involve only simple arithmetic, they are usually confusing to a beginner. Practice in working problems in essential to obtain a clear understanding of the various approaches to such calculations. Some of the basic types of calculations include.

1. Determination of equivalent weight of reagents from their behaviour in reactions.
2. Conversion of weights of substances to millimoles or milliequivalents and the reverse.
3. Conversion of concentrations expressed as weight of solute per unit volume of solution, or specific gravity and percentage by weight, to the system of molarity and normality.
4. Conversion of concentration expressed as molarity or normality to weight per unit volume.
5. Use of normality in the analysis of substances by titration.
6. Normality of a solution resulting from a mixture of components.
7. Normality of a solution resulting from dilution.

Examples illustrating the calculations required for 1, 2, 3, and 4 and have been given in the preceding sections. An example for 5 has been given for neutralization reactions and will be extended to other types of titrations in the chapters that follow. Examples for 6 and 7 are:

Example : What is the normality of a solution that results from the mixing of 50.00 ml. of 0.2400 *N* NaOH with ml. of 0.1280 *N* NaOH ?

Find the total milliequivalents present and divide by the total volume.

$$\frac{(50.00 \times 0.2400) + (25.00 \times 0.1280)}{75.00} = 0.2026\ N$$

Example : How much water should be added to 100.0 ml. of 0.2500 N HCl to give a 0.2000 *N* solution ?

Here the total number of milliequivalents present before dilution must equal the total number present after dilution,

$$(100.0 \times 0.2500) = (100.0 + x)\ 0.2000$$

and x, the amount of water added = 25.00 ml.

In both of these examples it must be assumed that the total volume is exactly the sum of the added volumes.

CHAPTER 5

General Remarks on Volumetric Analysis

Volumetric (titrimetric) analysis is a very widely used method. Watters has given a very lucid discussion of the general background of titration metric analysis. Chalmers has written guidelines on the development and publication of new titrimetric methods.

Advantages

Watters states that titrimetry has the following inherent advantages over the direct instrumental evaluation of a physicochemical property:

1. Accuracy,
2. Reliability,
3. Base and rapidity of performance,
4. Specificity through adjustment of conditions,
5. Adaptibility to a large variety and to a wide range of sizes of samples, and
6. Simplicity of interpreting the results.

Betteridge states that an important advantage of titrtimetry is its versatility, nearly all elements and many organic compounds can be determined by titrimetric procedures. Betteridge also comments the simplicity of the classical titrimetric technique. A first grade burette costs about Rs. 100. No other apparatus or instrument of comparable accuracy can be bought for this price. Simply routine breakdowns such as a jammed stopcock or a blocked tip can be dealt within a few minutes even by an unskilled technician.

Watters states that one of the most important advantages of titrimetry over direct instrumental analysis is the fact that it is an absolute method. With volumetric method, no related series of knowns are needed and an entirely different species can be used to evaluate the reaction capacity of the reagent.

In considering a volumetric method; three aspects should be considered–characteristics of the titrimetric method, characteristics of the titration reaction of characteristics of the titrant.

General reqirements of a method. Chalmers suggests that a titrimetric method should:–

(*a*) be accurate and precise.

(*b*) be simple to use.

(*c*) work under a wide range of conditions.

(*b*) fulfil a definite need.

(*c*) be rapid and

(*f*) have practical applications.

The term "accurate and precise" means that the results should be correct and reproducible within the limits set by any random errors in measurement, and it is imperative that the errors be kept as small as possible. "Simple to use" and "work under a wide range of conditions" means that there should be as few procedural steps involved as possible. The tolerance of pH range, and the amounts of other compounds permitted to be present should be as wide as possible. "Rapid" means that the reaction should proceed more rapidly than titrant can be added to the system.

Chalmers also states that unless a method being developed promises to fulfil these criteria, it seems scarcely worthwhile pursuing it. If it does show promise, proper investigations of the conditions under which it should be used must be carried out and a through assessment of its scope of application and reliability should be made.

General requirements of a volumetric reaction. Ayres states that for application of a titrimetric method, the chemical process involved should consist of a single, rapid stoichiometric reaction between the substance titrated and the standard solution used as the titrant.

Watters has suggested the following requirement or a direct titration reaction.

1. it should go essentially to completion
2. It should be exceeding rapid.
3. It should be stoichiometric.
4. It should be free from side reactions.
5. It should be specific for the substance to be determined.
6. A means of precise evaluation of the equivalence point should be available.

General requirements of a titrant. Chalmers suggests the following specifications for a good titrant.

1. It must have a well defined composition.
2. It must be easily soluble in the chosen solvent so that solutions covering a wide range of concentrations can be easily prepared.
3. Solutions of the titrant should either be stable, or they should be easily standardized.

General factors influencing volumetric method.

(*i*) **Ionic strength.** The ionic strength can affect both the velocity and the quantitativeness of a chemical reaction, the effect being largely a function of the charges on the reacting species. In a titration there must inevitably be a change in ionic strength in the course of the reaction, but fortunately, most analytical procedures require the use of concentrations at which activity coefficients change relatively little with variation in ionic strength, so it is unlikely that the ionic strength will be a serious factor.

Changes in the concentration of the reactants cause a change in the reaction velocity. This is seen in the decrease in the reaction velocity near the equivalence point, which has a practical utility. The entering stream of titrant reacts immediately in the early part of a titration, but disperses sluggishly near the end-point, so that at the point of equivalence, the indicator has changed colour. It is therefore convenient to keep reactant constant during the preliminary investigations. Later the concentrations can be varied, especially with a view to finding the lowest concentrations that can be used satisfactorily.

(*ii*) **Temperature.** A change in the temperature influences :he reactions velocity and the equilibrium constant, and sometimes even the stoichiometry of the reaction. Sometimes a higher temperature is used deliberately to make a slow reaction more practicable and sometimes a very low temperature is used to make an interfering reaction so slow that its effect is nullified (kinetic masking).

Ideally, room temperature should be the optimum temperature, but it is necessary to investigate the effect of changing the temperature because 'room' temperature may vary from a few degrees in winter in Kashmir to 40° or more in summer in Delhi. High temperatures are undesirable because they may cause unfavourable effects such as decrease in stability of reactants, thermal composition, reaction with solvent, increasing probability of occurrence of side reactions.

(*iii*) **Stability of reagents.** It sometimes happens that through a titrant may be stable in aqueous solution when it is being stored, it may become less stable under the conditions used for the titration, especially if these are particularly drastic in terms of acidity or alkalinity.

(*iv*) **End-point sharpness.** The end point of the titration is detected by an indicator which shows a marked change in some property of the titration system, this change should coincide with the stoichiometric point of the titration reaction.

The accuracy of a titration will be partly determined by the volume of titrant required to traverse the end-point, and know how accurately the end-point can be located within that volume. Sometimes, the equivalence point may not exactly correspond to the end point, and correction may be necessary; such a correction may be calculated or empirical.

Results and applications. Once the method has been worked out, it is necessary to validate it by standardization, and by application to standard samples containing the species of interest. The standardization is customarily done by preparing a standard solution of known concentration and titrating equal portions of it. This is not a sound practice. First, all measurements are made by volume, it is therefore necessary to calibrate all apparatus used, so that the personal error may be assessed. Secondly, once the first titration has been done, the tendency is to add rapidly almost the required volume of titrant to the second sample and only the last few drops are added slowly. The result is a variable drainage error, if fractions of a division are estimated in reading the burette, there is an unconscious urge to make the readings agree as closely as possible. It is much more reliable to make up the standard solution by weight and to take different weight fractions of it. Each titration must then be done more or less individually without reference to the others, and the error in the amount of sample taken will be virtually eliminated. If a sufficiently wide range of sample weights is taken a plot of 'taken' against 'found' will reveal any bias in the results, and the existence of any 'blank' correction. Repetition of the procedure in the presence of the other elements expected to occur in any application will reveal the extent of any interference by these species. The concentration ratios used must be realistic.

Finally, the method must be applied to standard samples of the materials for which it is suitable, and the results compared with those obtained by the best previously existing methods.

CHAPTER 6

Evaluation of Analytical Data

ACCURACY AND PRECISION

The numerical value obtained in every scientific measurement contains certain degree of uncertainty. In very accurate work the uncertainty may be very small, but it is still present. This uncertainty is called the error of the measurement. Accuracy relates to the closeness of approach of a single measurement, or of the average of a series of measurement, to the true value. Accuracy may also be defined as the concordance between the measured value and the true or most probable value. The true value cannot be measured exactly, however, in many cases it can be estimated very closely. By using calibrated equipment, performing the work very carefully, executing a very large number of measurements, and then applying statistics to the results, reasonable approximations of the true value can be obtained. Mostly, the true value is estimated from the results of different analysis in different laboratories.

Precision

It described the closeness of approach of replicate results to a common value. Repeated measurements of the same quantity will usually not be identical, but will scatter around some common value. Precision describes the reproducibility or scatter of a series of measurements of result. Precision may also be defined as the concordance of a series of measurements of the same quantity.

Both accuracy and precision have specific meanings and implications, so they should not be used interchangeably or carelessly. To illustrate exactly what is ment by the two terms, accuracy and precision, let us examine two sets of data for the weight of a crucible. The data in set *A* were obtained by five different students each weighing the crucible on his own analytical balance, each using, therefore, a different set of weights. The second set *B* were obtained by weighing a sixth student weigh the crucible five times on his analytical balance, thus these data were obtained on the same equipment. The crucible is then weighed by a skilled analyst using ASTM-II-S-I weights, and applying all appropriate corrections to the weight obtained. He finds a value of 9.2474 g as the average of ten replicates, with no statistically significant trends or deviations in the individual weights or in the means. Thus, it is fairly safe to assume that this value is sufficiently close to the true value to be considered the correct weight of the crucible.

	Set A	*Set B*
Weight of the crucible	9.2463	9.2483
	9.2480	2.9481
	9.2477	9.2484
	9.2489	9.2480
	9.2455	2.2483
Average, weight, g	9.2473	9.2482

Let us now examine the data in sets *A* and *B*. The weights obtained in set *B*, by the one operator using the same equipment are less scattered, much more closer to each other than those found in set *A*. The average of weights obtained by the five students (Set *A*) is closer to the true value than is the average of weights in set *B*. These data reveal an interrelation between accuracy and precision that is very important. Precise data need not necessarily be accurate. A constant and reproducible error can be contained in each piece of data. The data in set A are misleading, for it is highly unlikely that poor precision will accompany good accuracy. Precision always accompanies accuracy, but a high degree of precision does not imply accuracy. This is illustrated by an example.

Example 6.1. *A substance was known to certain 49.06% of given constituent A. The results obtained by two observers using the same substance and the same general technique were :*

Solution. Observer (1) 49.01 ; 49.21 ; 42.08. Mean = 49.10%

Relative mean error $= (49.10 - 49.06)/49.06$

$$= 0.08\%, \left[\frac{0.04}{49.06} \times 100 = 0.08\%\right]$$

Relative mean deviation^{+} $= (0.09 + 0.11 + 0.02)/3 \times 100/49.10$

$$= 0.15\%$$

Observer (2) 49.40 ; 49.44 ; 49.42. Mean = 49.42%.

Relative mean error = (49.42–49.10)/49.06

$$= 0.73\%. \left[\frac{0.32}{49.06} \times 100 = 0.73\%\right]$$

Relative mean deviation* = (0.02 + 0.02 + 0.00)/3 × 100/49.42

= 0.03%.

The analyses of observer (1) were therefore accurate and precise ; those of observer (2) were unusually precise, but less accurate than those of observer (1). Some small source of error appears to be present in the results of (2).

Method for the expression of Accuracy

The usual expression of accuracy is merely the difference between the piece of data (or the average of a set of data and the true value). This is known as the absolute error or the mean error. Generally, the absolute error by itself is of low information and of little use unless one knows the magnitude of the quantity being measured of calculated. Thus saying that there is 3.0 mg low error in weighing sample, while having some value, takes on a much greater significance if we know what the weight of the object is. It makes a difference whether the weight is 30 g or 30 mg. The error is usually negligible in 30 g, but is very significant in 30 mg.

Thus the accuracy of a measurement is usually expressed in relative terms, as a fraction of the true value. It is expressed in % or in parts per thousand (p.p.t.)

Relative accuracy/Relative error

= Absolute error/True value × 1000

Relative accuracy in the data *A* : –0.0001/0.9474 × 1000

= – 0.01 ppt

Relative accuracy in the data *B* : + 0.0008/9.2474 × 1000

= + 2.086 ppt

It should be pointed out that as a general rule, relative accuracy and relative precision expressions are rounded off to the first decimal place. Note that smaller the numerical value of the relative error (disregarding the sign) the more accurate the data. Thus the data in set *B* are less accurate than in set *A*.

Method of the expression of precision

There are many different ways to express the precision of a set of a data. Of these, only a few have any statistical basis, causing these expressions now to be recommended. Today these are the most frequently encountered terms. A couple of the terms below are listed for the historical purpose.

The Range, W. This is merely the numerical difference between the highest and the lowest value of a set of results. For the data in Set A, $W = 9.2489 - 9.2455 = 0.0034$ g, use of the range is not particularly informative since it tells nothing about the distribution of the data. The data could be distributed uniformly between the two limits or they all could be clustered around one value, with just a single datum way off. The range is most frequently used for very small set of data, i.e., two or three items. The range becomes less informative as the number of pieces of data in a set increases. The range is more informative about the precision of a set of data than the average deviation when there are fewer than eight pieces in the set.

The other methods for expressing precision are based on the absolute difference between each particular piece of data and the average of the set. This difference is known as the deviation of a particular piece of data. A deviation has no sign, being obtained by subtracting the datum from the average of the set. Thus for any given measurement, X_i, in a finite set of data whose average arithmetic mean is m, the deviation is $= [X_i - m]$.

The mean, of course is $\Sigma_i, X_i/n$

Where n is the number of pieces of data in the set. The deviation in the sets A and B are given below:

Deviation for set A	*Deviation for set B*
0.0010	0.0001
0.0007	0.0001
0.0004	0.0002
0.0016	0.0002
0.0018	0.0001

The average Deviation

Before 1960 the average deviation was very popular for expressing the precision of a set of data. The only reason for its wide spread use was that it is relatively easy to calculate. Statistically, the average deviation has no importance, for it does not convey any useful information and these days it is rarely used. The average deviation is nothing more than the average of all of the deviations for the members of the set of a data taken without regard to sign, $a = \Sigma_i(X_i - m)/n$.

Thus for the data in set *A* the average deviation,

$$a = (0.0010 + 0.0007 + 0.0004 + 0.0016 + 0.0018)/5 = 0.0011$$

and for the data in set *B* the average deviation is

$$a = (0.0001 + 0.0001 + 0.0002 + 0.0002 + 0.0001)/5 = 0.00014.$$

Relative Mean Deviation

The relative mean deviation is the mean deviation divided by the mean. This may be expressed in terms of percentage or in parts per thousand. Thus the relative mean deviation for the set of data *A* (Relative mean deviation = $a/m \times 1000$); $0.0011/9.2473 \times 1000 = 0.11$ parts per thousand.

Thus for the data in set *B*, the relative mean deviation is $= 0.0014/9.2482 \times 1000 = 0.015$ parts per thousand.

The Standard Deviation

It is a meaningful statistical term used to express precision. If one should happen to have a set of data that contains an infinite number of pieces, then a would be the parameter used to described the standard deviation of the members of the set. Normally, however, scientists have only finite set of data, and infact, in analytical chemistry, usually only a very small set of data (2,3,4 or 5 members). Thus the symbol used for the standard deviation is *S*, indicating that we are treating only a limited set of data. From the way that the standard deviation is formulated, more weight is given to pieces of data having large deviations. This is appropriate since they occur less frequently. The standard deviation is defined as the square root of the sum of the squares of the deviations divided by one less than the number of pieces of data in the set namely,

$$S = \sqrt{\Sigma_i (X_i - m)^2/n - 1}$$

The standard deviation for a finite set of data is obtained by dividing by one less than the number of pieces of data since this denominator represent the number of degree of freedom the set of the data possesses which are available for asserting error. Since the mean rather than the true value is used in calculating the deviations, and it is a property of the set that $\Sigma (X_i - m) = O$, when one has selected n—1 values of the deviation, the last deviation cannot be arbitarily selected but is predetermined. Thus when calculating *S* for any set of the data possessing the mean, *m*, only n–1 degrees of freedom, show the effect or error.

The square of the deviations for data in set A and B

Set A	*Set B*
$(X_i - m)^2$	$(X_i - n)^2$
100×10^{-8}	1×10^{-8}
49×10^{-8}	1×10^{-8}
16×10^{-8}	4×10^{-8}
256×10^{-8}	4×10^{-8}
324×10^{-8}	1×10^{-8}
Sum = 745×10^{-8}	11×10^{-8}

Consequently the standard deviation for the data in Set A

$$S = \sqrt{745 \times 10^{-8}/4}$$

$$= 1.4 \times 10^{-3}\ \text{g} = 0.0014\ \text{g}$$

and the standard deviation for the data in Set B

$$S = \sqrt{11 \times 10^{-8}/4}$$

$$= 2 \times 10^{-1} = 0.0002\ \text{g}$$

The smaller the value of S, the more precise the is data. Note that the numerical value of the S is greater than of the average deviations for the data in both of the sets. This is due to the increased weight given to the values with larger deviations.

The variance S^2. For certain statistical tests it is more convenient to use the square of the standard deviation, instead of the standard deviation. This term is called the variance, and it is defined as the sum of the square of the deviations divided by one less than the number of pieces of the data or

$$S^2 = \Sigma_i (X_i - m)^2/n{-}1$$

The variance of the data in set A is 1.96×10^{-6}, and the variance of the set of the data in $B = 4 \times 10^{-8}$

Relative Standard Deviation (Coefficient of Variation)

As pointed out above, when discussing accuracy, the absolute value of the standard deviation is of limited value. Usually the value of the mean of the data set is given along with the standard deviation which does enable one to estimate the relative magnitude of the S. When the S is expressed as a fraction of the mean of the set of data, a better feeling for the proportionality of the error is obtained. The relative S is nothing more than the S divided by the mean of the set. Again is it expressed in parts per thousand in order to avoid the possible confusion by the use of % values. Relative Standard Deviation = $S/m \times 1000$

For the set A ; 0.00014/9.2473 × 1000 = 0.15 ppt

For the data in set B = 0.0002/9.2482 × 1000 = 0.02 ppt

Thus, once again it is found that the data in set B is more precise than those in the Set A. The smaller the value of the relative standard deviation, the more precise the data set is.

Probable Deviation

The probable deviation is only used as a means of expressing precision when there is a very large number of pieces in the set of data (i.e., more than 100). The probable deviation is the most frequently occuring deviation. When all of the positive deviations are arranged sequentially, the probable deviation is that deviation so chosen that there are as many deviations with magnitudes greater than it as positive deviations with a magnitude smaller than it. The same value should be obtained by performing the identical manipulation on all the negative deviations. Statistically it is readily obtained from the standard deviation.

Probable Deviation (p) = $0.67 \times S$

Confidence Limits or Interval

This is a statistical term used to define the distance on either side of the mean value of a set of data, within which one may expect to find, with stated probability, the true value. Thus the 95% confidence interval or limit is the distance on either side of the average within which the true value will be found with 95% of the time. Going to 99% confidence limits, increases the probability of including the true value, but naturally it must be a broader region than the 95% confidence limits.

Confidence limits are defined from the t test (giving below) as confidence limits = $\pm S/\sqrt{n}$, where n is the number of pieces in the set of the data and t is the number depending upon the number of pieces of data and the confidence level desired. Value of t are tabulated in statistics books. A selected portion of such a Table is given under the t test for 1 – 10 pieces of data and 90%, 95% and 99% confidence level. For most analytical work 90 or 95% confidence is sufficient, although for very accurate work, 99% confidence levels are used.

For the data in set A at 95% confidence, the confidence interval is = $2.78 \times 0.0014/\sqrt{\ } = \pm 0.0017$ g

Therefore the value lies within 9.2473 ± 0.0017 g at 95% confidence level. For the data in set *B* for 95% confidence level = $\frac{2.78 \times 0.002}{\sqrt{5}}$ = ± 0.002 g and the true value lies within 9.2482 ± 0.0002 g at 95% confidence level. (Since we have already said that the error in set *B* was not a random error but a systematic one, it is not to be expected that the true value in the case of set *B* will be statistically predictable)

Note that the 99% confidence interval for set *A* is 0.0029 g and that for the set *B* is = 0.0004 g.

Classification of Errors

In looking over the errors that are apt to occur during a chemical analysis, one sees that they are of two major types, those of known origin that can be corrected for and those whose origin is unknown. On the basis, errors are generally divided into two categories–determine or systematic errors and indeterminate or random errors.

Determinate Errors

The most important of these are personal errors. In a preponderance of cases where a gross error exists in the results obtained by an established method of analysis the cause lies with the experimenter. He may have failed to take down the correct value for a weighing, having either misread his weighing or transposed the correct numbers. He may have erred in the reading of a volume. The beginning student in quantitative analysis may incur errors of a manipulative nature, failing to transfer materials completely from one container to another or allowing a sample to become contaminated in some way. Failure to introduce the proper amount of a reagent in the course of an analysis must also be classed as a personal error.

These very common errors can, of course, be easily avoided by the exercise of care. For example, conscientious verification of every experimental reading is excellent practice, the time required for this is modest indeed and will serve to eliminate errors of this sort. One personal error that should always be guided against prejudice. Where an estimation is involved, there is natural tendency for the experimenter to choose that value which is most favourable to him. Intellectual honesty is the keystone of all sciences; it is essential for the attainment of significant analytical results.

Instrumental Errors

Instrumental errors are those errors that are attributed due to imperfections in the tools with which the analyst works. The tolerances of the chemists weights may be the potential source of an instrumental error ;

mishandling of analyticarl weights will often cause them to differ significantly from their nominal values. Volumetric equipment—burets, pipets, and volumetric flasks-frequently deliver or contain volumes slightly different from those indicated by their graduations; for analytical work of the highest order, these variations must be measured by calibration and taken into account; otherwise they become a source of instrumental error.

Method Errors

Analytical procedures are also subject to limitations ; these give rise to errors that can be traced to the method. For example, even the least soluble of substances has a finite solubility. In gravimetric analysis the chemist is confronted with the problem of isolating the element to be determined in the form of a precipitate of the greatest possible purity. If he fails to wash it sufficiently, the precipitate will contaminated with foreign substances and have a spuriously high weight. On the other land, excessive washing may lead to the loss of weighable quantities of the precipitate; this will have opposite effect upon the results. It should be clear that an error will result in either case, and that there is a limit of accuracy that can be attained by such an analysis.

A method error that is frequently encountered in volumetric analysis arises from the fact that a volume of a reagent in excess of the theoretical is required to cause the change of colour that signifies completion of the reaction. As a result the ultimate accuracy of an analysis is often limited by the very phenomenon that makes the determination possible.

The undiscovered presence of interfering contaminants in either sample or reagents can lead to serious determinate errors in the results yielded by an analysis. Errors inherent in a method are probably the most serious of determine errors in that they are the most likely to remain undetected.

Constant Errors

Those errors whose magnitude is independent of the quantity measured are called constant errors. For a given analytical procedure, a constant error will be more serious as the size of the quantity measured decreases. For the purpose of illustration we will suppose that the washing of a slightly soluble precipitate with 100 ml of water is called for, and that 0.5 mg is lost during this process. If 500 mg of precipitate are so treated, the relative error due to this cause will be $-(0.5 \times 100/500)$ or -0.1% present, which for most purposes is tolerable. The loss of this quantity from 50 mg of precipitate will be responsible for a relative error of $-(05 \times 100/50)$ or 1%, which may no longer be acceptable.

The amount of reagent required to bring about the colour change in a volumetric analysis is also an example of constant error. This volume, usually small, remains the same regardless of the total volume of reagent required. Again, the relative error will be more serious as the total volume decreases. Clearly, one way of minimizing the effect of constant errors is to use as large a sample as is consistent with the method at hand.

Proportional Errors

Proportional errors, on the other hand, depend in absolute magnitude upon the quantity measured and increase or decrease in proportion to its size. The presence of interfering contaminants, if not eliminated in some manner, will lead to an error of the proportional variety. For example, a method widely employed for the analysis of copper involves the reaction of the cupric ion with potassium iodide ; the quantity of iodine produced in the reaction is then measured. Ferric ion, if present, will also liberate iodine from potassium iodide. Unless steps are taken to prevent this interference, the analysis, will yield erroneously high results for the percentage of copper since the iodine produced is a measure of the sum of the copper and iron in the sample. Thus the magnitude of this error is fixed by the extent of iron contamination, and will produce the same relative effect regardless of the size of the sample taken for analysis. If the sample size is doubled, e.g., the amount of iodine liberated by both the copper and the iron contaminant will be doubled ; and while the absolute error will also undergo a two fold increase the relative error will remain unchanged.

Detection and Minimization of Determinate Errors

From the standpoint of magnitude, determinate errors are more important in analysis, and in some instances are difficult to detect. They can effect a single result, a series of results, or an entire method of analysis depending upon their nature and the phase of the analysis in which they occur. A personal error, such as the misreading of a weight while weighing a sample, could result in a spurious value for that sample alone; the same error incurred during the preparation of a reagent might cause the propagation of a proportional error throughout all analysis undertaken with that solution. In the first instance the divergence of the result from the others in the series would possibly provide a clue that the error had been incurred. In the absence of knowledge of the approximate value to be expected, the error in the latter case could well pass undetected. An error of this sort would be revealed by repetition of the analysis with an other reagent or by a check analysis with the original solution against a sample of established purity.

Instrumental errors attributable to the discrepancies from the nominal values in analytical weights and the volumes contained by volumetric flasks or delivered by pipets and burets are generally quite small in new equipment of modern manufacture ; with having use of mishandling, however, they can acquire serious magnitude. To eliminate errors from the sources, the periodic calibration of such equipment is a necessity.

Errors of method are the most serious, since they effect all results obtained by the method. The detection of errors inherent in analytical scheme may take any of several courses; some of those that are generally applicable merit discussion.

Analysis of Standard Samples

A method may be tested by employing it for the analysis of synthetic samples whose overall composition approximates that of the substance for which the analysis is being considered. Great care is excercised in the preparation of these standard samples to insure that the concentration of the constituent to be determined is known with great accuracy.

Proportional as well as constant errors in a method may be detected through the use of standard samples. Studies aimed at the development or improvement of analytical procedures often use this approach; but the preparation of samples whose composition resembles that of a complex natural substance frequently is difficult. Furthermore, the difficulties are multiplied by the requirement that the exact concentration of one of the constituents be known as a result of the method of preparation. In some cases problems can be so imposing they prevent the application of this procedure.

Independent Analysis

Closely associated with the foregoing is the parallel analysis of a sample by a method of established reliability that is dependent of the one under investigation. This is of particular value where sample of known purity are not available ; it also has the effect of supplying a frame of reference by which the utility of a new method may be judged. In general, the more widely the independent method differs from the one under study, the more suitable it is for this purpose ; this trends to eliminate the possibility that some common factor in the sample will affect both equally and thus remain undetected.

The National Bureau of Standards should be mentioned in this connection since it has made available for purchase a fairly large number of common substances that have been carefully analyzed for one or more constituents. These are very valuable for the testing of analytical procedures for accuracy.

Blank Determinations

Constant errors affecting physical measurements can frequentiy be evaluated with a blank determination ; this consists of performing all steps of the analysis in the absence of a sample. The result is then applied as a correction to the actual measurements.

Blank determinations are of particular value in exposing constant errors that are due to causes such as the presence of interfering contaminants in the reagent employed in the analysis and the slight deficiencies often encountered in the indicator of titration end points ; the latter are referred to the indicator corrections.

Variation in Sample Size

In detecting errors in an analysis, applying the procedure to different-sized samples of the same material is sometimes helpful. The presence of constant errors will become obvious in the discrepancies appearing among the results of the several analyses. For example, the data in Table 6.1 were obtained from various sample sizes of a silver alloy containing exactly 20% silver. In each of these analyses, there was an error of 2 mg in the weight of silver found. This resulted in a low values for the percentage of silver. The differences in the percent silver found between samples, however, becomes less as the sample size is increased, and the data appear to be approaching a constant value with the larger samples.

Table 6.1. Effect of Constant Error of 2 mg on the Analysis of a Silver Alloy

Weight of Sample (in grams)	*Weight of Silver Found (in grams)*	*Silver Found (percent)*
0.2000	0.0378	18.90
0.5000	0.0979	19.54
1.0000	0.1981	19.81
2.0000	0.3982	19.91
5.0000	0.9980	19.96

By way of contrast, proportional error goes undetected by this procedure. In the same way are plotted data from the analysis of various samples of the same material by a method containing a proportional error of 5 parts per thousand. The data are given in Table 6.2. The plot in this case will be a straight line ; clearly, without knowledge of the time percentage silver in the alloy, the existance of such an error would go unnoticed.

Table 6.2. Effect of Proportional Error of 5 parts per Thousand on the Analysis of a Silver Alloy

Weight of Sample (in grams)	*Weight of Silver Found (in grams)*	*Silver Found (percent)*
0.2000	0.0402	20.10
0.5000	0.1006	20.12
1.0000	0.2009	20.09
2.0000	0.4021	20.11
5.0000	1.0051	20.10

Indeterminate Errors

Indeterminate errors are those errors that are caused by unknown or uncontrolled factors. Once the source of a random error is known, it becomes a systematic error. This class should be truely random in nature, causing results that are high just as frequently as they are low. Random errors cannot be corrected, because they are due to normal fluctuations in the behaviour of the equipment or of the experimenter. The equilibrium position of a balance will vary so slightly from weighting to weighing, the movement of a meter will cause the needle of a pH meter or a spectrophotometer to give slightly different readings each time, an experimenter will read the minicus on a burete a bit different each time, drainage through a chromatographic column packing will be slightly different each time causing different R_f values. These are the normal, daily fluctuations in performance and occur in a random fashion.

Indeterminate errors can be treated by statistical methods and it is only indeterminate errors for which statistical treatment is valid.

The Normal Law of Distribution of Indeterminate Errors. The distribution of indeterminate errors is expected to follow the normal distribution (Gaussian) law. This law is given by the equation :

$$Y = \frac{1}{\sigma(2\pi)^{1/2}} e^{-(X_i - m)^2/2\sigma^2}$$

where Y is the frequency of occurrence of a deviation of a particular magnitude, σ is the standard deviation, X_i is the experimental datum. m is the mean of the set, and this $X_i - m$ is the deviation of the value. This experimental relationship between a deviation caused by a random error and the frequency with which it occurs, strictly speaking is valid only for extremely large numbers of pieces of data. For smaller sets of data, such as are normally encountered in analytical laboratories, the scatter of deviations will not follow this relationship. By considering these data as only a few points, which are a part of much large set of data which does obey a Gaussian distribution, statistics and statistical tests can be applied successfully to small set of data. A graph of the frequency occurence of a deviation gives the familar bell-shaped normal distribution curve (Fig. 6.1).

The standard deviation occurs at the inflection point rising decending portions of the curve. Sixty eight percent of the area of the normal distribution curve lies between + σ and –σ. This means that two out of the three pieces of data deviate from the mean by only one standard deviation or less.

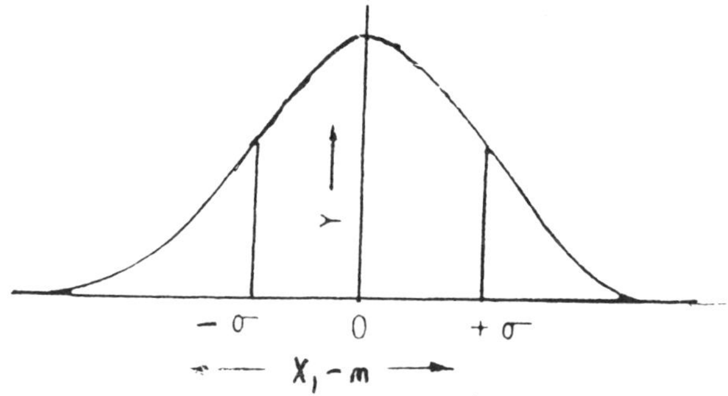

Fig. 6.1 The Normal Distribution Curve.

Two features of the normal distribution curve should be pointed out. First, the curve is symmetrical. That is to say, positive and negative deviations occur with equal frequency (probability), second small deviations occur more frequently than large deviations, or to phase it another way, the larger the deviation, the less apt it is to occur. A corolary of this, however, is that although small deviations occur most frequently, large deviations do have a finite probability of occurring. This feature of the error curve takes on importance when deciding whether or not to reject a piece of data that is not very close to the rest of the data of the set.

Statistical Tests of Data :

The F Test. A problem frequently encountered in evaluating analytical data involves judging whether or not the precision of two sets of data is essentially the same. Do two different analytical methods give data with roughly the same precision, or is one method considerably less precise than the other ? The *F* test is used to compare the precision of two sets of data. *F* is the ratio so set up that it is always greater than one ($S_1^2 > S_2^2$). In the *F* test, after the value of *F* has been calculate from the two sets of data, the value is compared with that given in Table 6.3 of statistical *F* values. When the calculated *F* is larger than the tabulated value, the data in set 1 are indeed less precise than those in set 2, at the specified confidence level. Table 6.3 gives values for F at 95% confidence level. V_1 is the number of degree of freedom of data of the set having larger variance and V_2, therefore, is the number of degree of freedom of the set of the data having the smaller variance.

Table 6.3. Values of F at 95% Confidence Level

	V_1									
V_2	2	3	4	5	6	7	8	9	10	∞
2	19.00	19.16	19.25	19.30	19.33	19.36	19.37	19.38	19.39	19.50
3	9.55	9.28	9.12	9.01	8.94	8.88	8.84	8.81	8.78	8.53
4	6.94	6.59	6.39	6.26	6.16	6.09	6.04	6.00	5.96	5.63
5	5.79	5.41	5.19	5.05	4.95	4.88	4.82	4.78	4.74	4.36
6	5.14	4.76	4.53	4.39	4.28	4.21	4.15	4.10	4.06	3.67
7	4.74	4.35	4.12	3.97	3.87	3.79	3.73	3.68	3.64	3.23
8	4.46	4.07	3.84	3.69	3.58	3.50	3.44	3.39	3.34	2.03
9	4.26	3.86	3.63	3.45	3.37	3.29	3.23	3.18	3.13	2.71
10	4.10	3.71	3.48	3.33	3.22	3.14	3.07	3.02	2.97	2.54
∞	3.00	2.60	2.37	2.21	2.10	2.01	1.94	1.88	1.83	1.00

Consider the data obtained during a determination of the equivalent weight of a weak acid. In method A, phenolphthalein was used to detect the end point, whereas in method B, the glass electrode was used as the detection device. Which set of the data is more precise ?

	Method A	*Method B*
	122.4 g	122.0 g
	122.0	122.2
	122.6	122.6
	121.8	122.4
		122.5
Average	122.2 g/equiv.	122.5 g/equiv.

Using the equation,

$$S^2 = \Sigma_i (X_i - m)^2/n - 1$$

The variance S^2 for the data A is 2×10^{-1}, whereas the variance for the data obtained by method

$B = 5 \times 10^{-2}$ Applying the F test

$$F = S_1^{\,2}/S_2^{\,2} = 2 \times 10^{-1}/5 \times 10^{-2} = 4$$

From Table 6.3, for $V_1 = 3$ and $V_2 = 4$ we find the value of $F = 6.59$ at 95% confidence level. Since the calculated value of F is smaller than this we can conclude that the data obtained using phenolphthalein are just as precise as those obtained with glass electrode.

Example 6.2 *You are developing a new colorimetric procedure for determining the glucose content of blood serum. You have chosen the standard Folin-Wu procedure with which to compare your results. From the following two sets of replicate analysis on the same sample, determine whether the variance of your method differs significantly from that of the standard method.*

Solution.

Your method (mg/dl)	*Folin-Wu method (mg/dl)*
127	130
125	128
123	131
130	129
131	127
129	125
Mean (m_1) = 127	Mean (m_2) = 128

$$S_1^{\,2} = \Sigma \frac{(X_1 - m)^2}{7 - 1} = 50/6 = 8.3, \quad S_2^{\,2} = \Sigma \frac{(X_2 - m_2)^2}{6 - 1} = 24/5 = 4.8$$

$$F = 8.3/4.8 = 1.7$$

The variance are arranged so that F value is > 1. The tabulated F value for $V_1 = 6$ and $V_2 = 5$ is 4.95. Since the calculated value is less than this, we conclude that there is no significant difference in the precision of the two methods.

The Student Test. In 1908 Gosset derived what he called student's t test to take into account the smallest of the usual number of measurement, encountered in practise and variations in the number of pieces of data used to calculate the average. He defined t as :

$$\pm t = (m - \mu) \frac{n^{1/2}}{S} \qquad ...(1)$$

where m is the average of the set of data, μ can be considered as the true value, n is the number of pieces of set of data, and S the standard deviation for the set. Rearranging the above equation

$$\mu = m \pm t \frac{S}{n^{1/2}} \qquad ...(2)$$

From equation (2) we can see that the true value will fall within a distance of $\pm ts/n^{1/2}$ (the confidence level). How large this confidence level is well dependent upon the number of measurements taken, their standard deviation, and confidence level desired. If in practice the true value (known from some other source) falls outside this interval, we can conclude that there is a systematic error in the data in addition to the random errors. When this occurs, attempt should be made to scrutinize the analytical method to locate and eliminate the systematic error. Values of t are tabulated in most texts on statistics, and partial listing is given in Table 6.4.

Table 6.4. Values of t for Different Confidence Levels

Degree of freedom *V*	90%	95%	99%
1	6.31	12.71	63.66
2	2.92	4.30	9.93
3	2.35	3.18	5.84
4	2.13	2.78	4.60
5	2.02	2.57	4.03
6	1.94	2.45	3.71
7	1.90	2.37	3.50
8	1.86	2.31	3.46
9	1.83	2.26	3.25
10	1.81	2.23	3.17

If μ is considered to be known value for the percentage of a constituent in a sample (this true value must be obtained from some other source than the method being tested for) and t test can be used to evaluate the quantity of an analytical method, the performance of an experimenter, or the effect of changing parameters within the method. If μ is considered to be the average of one set of data and m the average of a second set of data pertaining to the same determination, one can use the t test to determine variations in the performance of personnel, equipment and the method.

The t test consists of calculating the value of t by means of equation (1) and comparing it with the statistically obtained value of t in tables. This is done at a particular confidence level decided upon by the judge of the data. When the calculated t exceeds the statistical value of t, there is a real difference (a systematic error) between the mean and the value. When the calculated t is less than the tabulated t, the difference between the mean and the true value is acceptable statistically (due only to random errors) and considered negligible (that is, the two values are identical.)

As an example let us apply the t test to the weighing data in set B on page 6 (for a crucible)

$$t = (9.2482 - 9.2474) \times 5^{1/2}/0.0002 = 4(2.24) = 8.96$$

From Table 6.4 we find that the value of t for 4 degree of freedom is 2.78 at the 95% confidence level and 4.6 at the 99% confidence level. Since our calculated value of t is much greater than either of these, there is a real error (systematic) in the data and search should be made for it since the values are not acceptable.

Example 6.3. *In a gravimetric determination of nickel with dimethylglyoxime, a student turns is the values, 3.681, 3.365, 3.635, 3.593%. The National Bureau Standard value is 3.557%. Does the average of these values, 3.632% of nickel, have a difference from the true value that is real or is it statistically possible to obtain such difference ?*

The calculated value for the standard deviation of the experimental result, S = 0.037%. There are four pieces of data in the set. Thus using equation (1) the student can calculate a value for t for the data.

$$t = (3.632 - 3.557)\,(4)^{1/2}/0.037 = 4.02$$

From Table 6.4 for three degree of freedoms and 95% confidence level the value of t is 3.18. Since his calculated t is greater than 3.18, the difference between the student's mean and the true value is real one, not merely due to statistical fluctuations of indeterminate errors. Note for 99% confidence level the tabulated value of t is 5.84. The student's calculated value is less than this, so that he should suspect a real difference (since the 95% t is exceeded) but have to conclude that it is statistically possible to have this large a discrepancy.

The comparison of results obtained by two different experimenters of from two different methods or under two different sets of conditions is a common valuable and necessary operation in analytical chemistry. One must be certain however, that the criteria used in many judgements have a sound statistical basis and not just whimsical. The t test provides an excellent basis for the comparison of the average value of two sets of data.

The procedure for comparing means using the t test is more complicated than that for comparing an average with the true value. Each of the two sets of data may have different number of pieces; each probably will have a different standard deviation so equation (1) is no longer applicable. A modification of it that compensates for these differences must be used, see equation (3)

$$\pm t = \left[\frac{m_1 - m_2}{s_{12}}\right]\left[\frac{n_1 \times n_2}{n_1 + n_2}\right]^{1/2} \qquad \text{...(3)}$$

Where m_1 is the average of the data in set 1 of n_1 pieces, and m_2 is the average of the data in set 2 of n_2 piecess. S_{12} is the standard deviation of any value in either set based on the data in both sets. It is computed according to equation (4)

$$S_{12} = \left[\frac{\Sigma_{11}(X_{i1} - m_1)^2 + \Sigma_{12}(X_{i2} - m_2{}^2}{(n_1 - 1) \times (n_2 - 1)}\right]^{2} \qquad \text{...(4)}$$

In fact, equation (4) represents a combination of the equations for the standard deviations of each set of data. This form of equation (4) is similar to that of equation for individual standard deviation determination of a single set of data.

In order to make a valid and meaningful comparison of two set of data, the precision of each set must be roughly comparable. So, before running the t test, an F test must be run to compare the precisions of the two sets of data. To illustrate the comparison of means using the t test following examples are given.

Example 6.4. *Student A determined the % of copper in a sample by using a partition chromatographic separation of the copper from the interference followed by EDTA titration to determine the number of milligrams of copper present. Student B determined the % of copper with the same unknown by extraction of copper-8-hydroxyquinolinate. A spectrophotometric measurement on the extract was used to determine the amount of copper. The results that each obtained are :*

	Student A	*Student B*
	37.35 *mg*, Cu	37.49 *mg*, Cu
	37.58	37.63
	37.17	37.87
	37.03	
	m_1 = 37.28 *mg*, Cu	m_2 37.66 *mg*, Cu

The true value is not known. The question arises: Is there a real difference between the two average obtained, or is this difference statistically acceptable, considering the different methods used?

In order to be able to compare two averages, the precision of the data in each set must be roughly comparable, i.e., the F test must be passed. The variance for the data obtained by student A is $S_1^2 = 565 \times 10^{-4}$ whereas by the student B it is 370×10^{-4}, $F = 1.53$. From Table 6.3, F is 19.16. Since our calculated F is less than that the tabulated value, therefore the precison of the two data are comparable. From equation (4) the standard deviation for the combined data sets can be calculated,

$$S_{AB} = \left[\frac{1695 \times 10^{-4} + 739 \times 10^{-4}}{3 + 2}\right]^{1/2} = 0.22$$

This can now be substituted into equation (3) along with other pertinent information,

$$\pm t = \frac{(37.66 - 37.28)}{0.22} \times \left[\frac{3 \times 4}{3 + 4}\right]^{1/2} = 2.26$$

Using Table 4 again, for five degrees of freedom (three in student A's data and two in student B's data) we find that the tabulated value of t at 95% confidence level is 2.57 and at 99% confidence level is 4.03. Since the value of t calculated above for these data is less than either tabulated value of t, we can say that the difference between the results obtained by these different methods is not meaningful and can be attributed solely to chance. Hence we can conclude that both methods will give us the same value for the amount of copper in the sample.

Example 6.5. *You are developing a new method for the determination of blood urea nitrogen (BUN). You want to determine whether your method different significantly from a standard one for analyzing a range of sample concentrations expected to be found in the routine laboratory. Following are two sets of results for a number of individual samples.*

	Your method	Standard method			
Sample	(mg/dl)	(mg/dl)	*Di*	$Di - \overline{D}$	$(Di - \overline{D}^{-2})$
A	10.2	10.5	—0.6	—0.6	0.36
B	12.7	11.9	0.8	0.5	0.25
C	8.6	8.7	—0.1	—0.4	0.16
D	17.5	16.9	0.6	0.3	0.09
E	11.2	10.9	0.3	0.0	0.00
F	11.5	11.1	0.4	0.1	0.01
			Σ 1.7		Σ 0.87
			$\overline{D}^{-} = 0.28$		

$$S = (0.87/5)^{\frac{1}{2}} = 0.42$$

$$t = 0.28/0.41 \times (6)^{1/2} = 1.6$$

Solution. The tabulated t value at 95% confidence level for five degrees of freedom is 2.571. Therefore, t calculated is smaller than the tabulated value, and there is no significant difference between the two methods at this confidence level.

Rejection of Data. Frequently, when a series of replicate analyses are performed, one of the results will appear to differ markedly from the others. A decision will have to be made whether to reject the result or to remain it. Unfortunately, there are no uniform criteria that can be used to decide if a suspect result can be ascribed to accidental error rather than chance variation. The only reliable basis for rejection is when it can be decided that some specific error may have been made in obtaining the doubtful result. No result should be retained in case when a known error has occurred in its collection.

Experience and common sense may serve as just as practical a basis for judging the validity of a particular observation as a statistical test would be. Frequently, the experienced analyst will gain a good idea of the precision to be expected in a particular method, and will recognize when a particular result is expected.

A wide variety of statistical tests have been suggested and used to determine whether an observation should be rejected. In all of these a range is established within which statistically significant observations should fall. The difficulty with all of them is in determining what the range should be. If it is too small, then perfectly good data will be rejected and if it is too large, then erroneous measurements will be retained too high a proportion of the time. In 1951, Dean and Dixon published an article dealing with statistics for small set of data, such as are normally encountered in analytical laboratory. Included in this paper is the Q-test for the rejection of data. This test has a sound statistical basis, and was designed for the use of only three to ten replicate analyses. The Q test is applicable when one and only one, result deviates from the rest of the set; it will not work if there are two results that are considerable distance from the average. Nor is the Q test applicable when all but one of the pieces of data in the set are identical. It will always reject data under these conditions e.g., given 3.413, 3.414, 3.414 mequiv/g as the experimentally obtained values for the capacity of an ion-exchanger resin, the Q test would permit rejection of the last value, 3.414 mequiv/g. Common sense tells us that 3.414 mequiv/g is a valid result, not divergent from the other two, and that it should not be rejected. The Q test states. When the value of Q calculated from the experimental data is equal to one or greater than the value of $Q_{0.90}$ tabulated for that number of observation, the suspected observation may be rejected at the 90% confidence level. The value of Q is calculated from the data by taking the distance of the suspect value from the nearest neighbour in magnitude, and dividing this by the range of the data, including the suspected value. That is $Q = (X_2 - X_1)/W$ (Fig. 6.2, $Q = a/W$). $Q_{0.90}$ is a statistical term computed at 90% confidence level. The use of it as a comparison in the Q test implies that one is correct 90% of the time in discarding a result that has calculated Q greater than $Q_{0.90}$. To phrase it differently, there is only a ten percentage chance that a result would be so divergent normally. Table 6.5 presents the values of $Q_{0.90}$ for three to ten pieces of data.

Table 6.5. Values of $Q_{0.90}$ for Three to Ten Observations

Number of observations	*Rejection Quitient, $Q_{0.90}$*
3	0.94
4	0.76
5	0.64
6	0.56
7	0.51
8	0.47
9	0.44
10	0.41

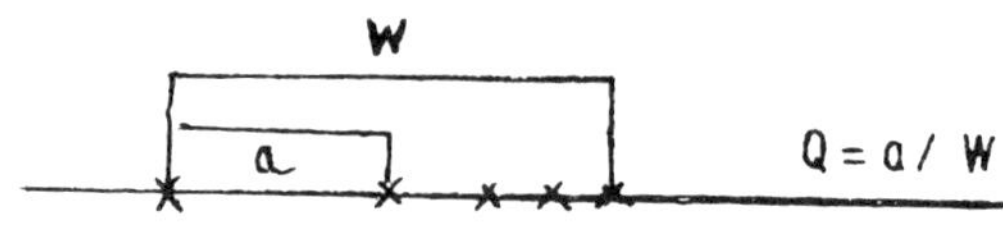

Fig. 6.2 Illustration of the calculation of Q.

Example 6.6 *The following set of chloride analyses on separate aliquots of a pooled serum were reported. One value appears suspect. Determine if it can be ascribed to accidental error. 103, 107, 114 meqllitre.*

Solution. The suspect result is 114. It differs from its nearest neighbour, 107, by 7meq/litre. The range is 1 4—103 or 11 meq/litre, The value of $Q = 7/11 = 0.64$. The tabulated value for four observations is 0.70. Since the calculated Q is less than the tabulated Q, the suspected number can not be rejected.

Example 6.7 *A student obtained the following results for the gravimetric determination of nickel by precipiation with dimethylglyoxime = 7.414%, 7.410%, 7.435%, nickel. The last value is noticiably higher than the other two, so the question arises concerning its validity and whether it must be included in the final average. The standard deviation and average value be much lower, if it could be discarded. The student is unaware of the any experimental reason for not including the value. Calculate the value of Q for the data,*

$$Q = 7.435—7.414/7.435—7.410 = 0.84$$

Solution. In Table 6.5, we find that the value of Q for three observations is 0.94. Since the suspect value gives of Q that is less than this, it is statistically possible to obtain it purely by chance, and thus it must be retained. Thus the average calculated % of nickel is 7.42% with a standard deviation of 0.113% nickel and a relative standard deviation of 1.8 ppt.

Example 6.8 *For the standardization of NaOH against primary standard potassium acid phthalate, the following values were obtained by a student of the normality of the solution 0.1014, 0.1012, 0.1015, 0.1021, 0.1012 N. All but 0.1021 N seem to cluster around a common value, so that 0.1021 N becomes suspect. Calculate the value of Q for the data.*

$$Q = 0.1021—0.1051/0.1021—0.1012 = 0.67.$$

Solution. From Table 6.5/$Q_{0.90}$ for five observations is found to be 0.64. Since the calculated Q is larger than that, the value of 0.1021 N should be discarded. Doing this, the average normality is 0.1013 N with standard deviation of 0.001 N and a relative standard deviation of 1.0 ppt. (Note that discarding the value of 0.1021 N. the average has dropped from 0.1015 N to 0.1013 N, the standard deviation is considerably better, down from 0.0004, with relative standard deviation decreased from 4 ppt).

For a small number of measurements (e.g., three to five) the discrepancy of the measurement must be quite large before it can be rejected by this criterion, and is likely that erroneous results may be retained. This would cause a significant change in the arithmetic mean, because the mean is greatly influenced by a discordant value. For this reason, it has been suggested that the *median* rather than the mean be reported when a discordant cannot be rejected from a small number of measurements. The median has the advantage of not being unduly influenced by an outlying value. In example one, the median could be taken as the average of the two middle values [=(106+107)/2=106]. This compares with a mean of 108, which is influenced more by the suspected number.

The following procedure is suggested for interpretation of the data of three to five measurements if the precision is considerably poorer than expected and if one of the observations is considerably different from the others of the set :

1. Estimate the precision that can reasonably be expected for the method in deciding whether a particular number actually is questionable.
2. Check the data leading to the suspected number to see if a definite error can be identified.
3. If possible, run another analysis. Agreement of the new result with the apparently valid previously collected will lend support to the opinion that the suspected result should be rejected.
4. If new data can not be collected, run a *Q* test.
5. If the *Q* test indicates retention of the outlying number, consider reporting the median rather than the mean for a small set of the data.

Method of Least Squares. The preparation of calibration curve is done in many chemical analysis. Ideally, most of the calibrations are linear. Thus, given a set of data that are presumed to obey a linear relation, the problem involved is to plot them and draw the best straight line through the points. This can be done, and frequently is done by inspection and intuitive placement of the transparent straight-edge on the points on the graph paper. Statistics provide a much sounder basis for drawing the straight line through the data points.

Let us assume that the data fit in the equation, $Y = mX + b$ where X is the concentration of the standard solution, and Y is the measured property of it, m is the slope of the straight line, and b have been determined, the proper straight line has been established, and it can be drawn through the points. It can be shown statistically that the best straight line through the series of experimental points is that line for which the sum of the squares of the deviations of the points from the value on the line (S) is minimum. This is known as the method of least square. If X_e and Y_e are the experimental data, and Y_e is the value of Y on the straight line for the parameter X_e, then the expression for the method of least square can be written ;

$$S = \Sigma_e (Y_e - Y_c)^2 = \Sigma_e [Y_e - (mX_e + b)^2], \qquad ...(1)$$

and the best straight line occurs when S is a minimum. By applying differential calculus to this, it can be shown that S is minimum when

$$m = \Sigma_n (X_e - \overline{X}) (Y_e - \overline{\gamma}) \Sigma (X_e - \overline{X})^2, \qquad ...(2)$$

$$b = Y - m\overline{X} \qquad ...(3)$$

where n is the number of points, $\overline{X}$ is the average of all the values of Y_e.

To illustrate, let us consider plotting the following data obtained in the determination of a metal by atomic absorption.

Concentration of Metal (X_e) μg/ml	Absorbance (Y_e)	$X_e - \overline{X}) (Y_e - \overline{\gamma})$	$X_e - \overline{X})^2$
0.000	0.000	0.580	1.030
0.390	0.228	0.214	0.390
0.780	0.476	0.022	0.055
1.560	0.888	0.173	0.298
2.340	1.267	0.922	1.759
$+\overline{X}$ =1.014	Y = 0.572	Sum =1.911	Sum =3.532

The average, $\overline{X}$, of the values of X_e (concentration of metal) is 1.014 μ g/ml and the average, $\overline{\gamma}$ of the value of Y_e (absorbance) is 0.572. The preceding table also includes the square of the deviation of X_e and their sum, and the product of the deviation of X_e times the deviation of Y_e and their sum. Thus from the equation (2). m=1.911/3.532 = 0.543 and from the equation (3), b = 0.572 - (0.543)(1.014) = 0.022, giving for the equation of the line through these data points, Y = 0.543 X = 0.022 Fig 6.3 shows the plot of the data and the line drawn through them according to the above equation.

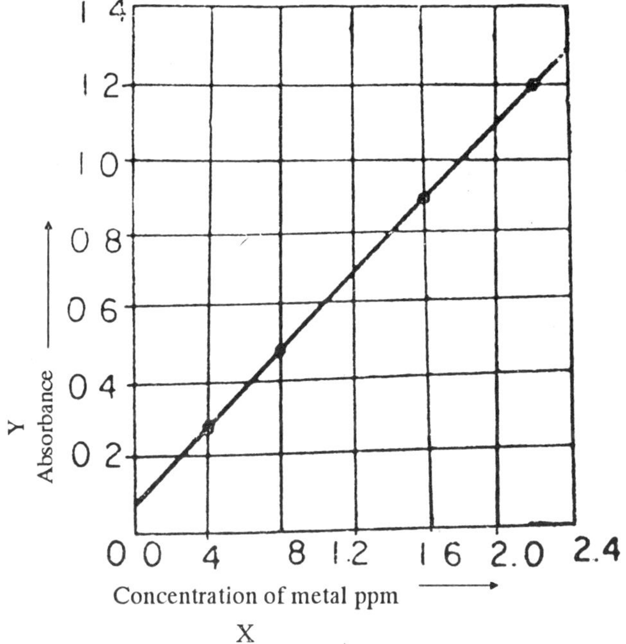

Fig. 6.3 Least Square plot for the atomic absorbance of a metal.

Significant Figures. A number used to express the results of a physical measurement or a calculation should contain sufficient digits to express the accuracy of the measurement or calculation. The general rule used in chemistry states that the least digit of a number, has certain uncertainty in it, while all of the other digits in the number contains no uncertainty. To illustrate, a sample is weighed on an analytical balance that has an accuracy ± 0.1 mg. To report its weight as 1.40633 g. is incorrect since the last decimal place is meaningless, because the balance is only capable of giving accurate information to the fourth decimal place. The weight should be properly written as 1.4064 g. Using the number of significant figures, the last digit, the 4 is uncertain by ±1 and this accuracy represents the capability of the balance. It is equally incorrect to express the sample weight as 1.406 g, for in weighing the sample, there is no uncertainty about the number of millgrams taken. Writing the weight with only three decimal place implies that there is. Full advantage of the accuracy and sensitivity of a measuring instrument should be taken and the results obtained with it should reflect this accuracy.

Let us try to determine the correct number of significant figures in a number, 0.4060 g. How many significant figures does it have ? Four. The zero to the left of the decimal point is merely to indicate the order of magnitude, to show that the weight is less than one gram. In no way it is a part of the number. The other two zeros are significant figures. As a general rule for numbers smaller than one, zeros proceeding the first non-zero digit are not significant, but merely indicate the order of magnitude of the number. Use of the exponential form for writing numbers eliminate confusion. Thus 0.004103 contains four significant figures, the three zeros proceeding the 4 are non significant. This can be seen more clearly when the number is written as 4.103×10^{-3}. When writing numbers larger than one, the use of the exponential form is essential for clarity. Zeros to the right of the last non zero digit of a number larger than one, can be either significant or not ; there is no way for the person reading the number to decide. Consider the number written as 4,510,000. How many significant figures does it contains ? three ? Fourth ? Seven ? When the number is written like this there is no way to judge. Some, perhaps all, of the zeros to the right of 1 may be merely to locate the decimal point. Some may be all, may be significant figures. In order to eliminate this confusion, the number should be written as say 4.510×10^{6}, which states clearly that there are four significant figures and still locate the decimal point.

It is customary at all intermediate stages of calculations to carry along one more significant figures than is proper. The calculation is rounded off to the correct number of significant figures only at the very end. This practice causes the error in the final result to be an accurate reflection of the final errors in the components. Only the final digit will contain a minimum of uncertainty.

In performing calculations one must be constantly alert so that the numerical result does not reflect an accuracy that is greatest or less than the least accurate number used in the calculation. When using computers, desk calculators or even with hand calculations, it is very easy to obtain long strings of digits. It is deceptive to believe these really reflect accuracy. In performing calculations we must be sure to dominate all meaningless, deceptive, excess digits from the results. The operational rules used for addition and subtration are different from those used for multiplication and division. In addition and substration, the number of decimal place limits the number of significant figures whereas in multiplication and division the number of decimal place is irrelevant. The result is restricted by the relative accuracy of the numbers involved in the calculation. The precision and accuracy of an experiment can never be increased by the computational process. Under certain conditions they may be decreased.

In addition and subtraction, the number of significant figures in the sum or difference can change from the number of significant figures of the components. Only the last digit of the sum or difference may contain any uncertainty. A sum or difference may contain no more decimal place than that component number having the fewest decimal place. Consider each of the following additions :

	1	2	3
	43.1	0.0025	10.414
	4.31	4.1167	0.0037
	0.431	5.9071	9.02
Sum	47.8	10.0263	19.43

In the first, all three numbers have identical significant figures, namely three, but 43.1 has the fewest decimal place and hence limits the number of significant figures in the answer. Only one decimal place can be used, so that 47.8 is the correct sum. In the second sum, all three numbers have the same number of decimal places, even though the first one has the fewest significant figures. The result can be expressed to four decimal places. Note that in this case the sum has one more significant figures than any number. In the third sum, all three numbers have different numbers of significant figures the sum is limited to two decimal places by 9.02. The number 0.0037 cannot be include in the sum because it lies beyond the limit of the accuracy of the least accurate of the numbers. Nothing can be added beyond the second decimal place.

The same general considerations apply to subtraction :

	1	2	3
	44.341	19.4197	4.96
	–4.4432	–19.4153	–0.00321
Difference	39.898	0.0044	4.96

The first difference is limited to three decimal place by 44.341. Note that in the second subtraction, the difference has only two significant figures, whereas each of the two components has six-a tremendous decrease in significant figures and hence, in relative accuracy of the difference. In the third set a subtraction cannot be made because the limit is at the second decimal place.

Multiplication and Division. In many measurements, one estimated digit that is uncertain will be included. This is the last significant figures in the measurement ; any digits beyond it are meaningless. In multiplication and division, the uncertainty of this digit is carried through the mathematical operations, thereby limiting the number of certain digits in the answer.

There are the same number of significant figures in the answer of multiplication or division as there are in the operator with the least number of significant figures. This is always the case, so you can say immediately how many significant figures will occur in the answer. We shall designate this limiting number as the key number. If there is more than one operator with the same lowest number of significant figures then the with the smallest absolute magnitude without regard to the decimal point is the key number.

Example.6.9. *Give the answer of the following operation to the maximum number of significant figures and indicate the key number.*

$$35.63 \times 0.5481 \times 0.05300/1.1689 \times 100 = 88.5470578\%.$$

Solution. The key number is 35.63. The answer is therefore, 85.55% and it is meaningless to carry the operation out to more than five figures (the fifth figure is used to round off the fourth). The 100% in this calculation is an absolute number, since it is used only to move the decimal point, and it has an infinite number of significant figures. Note that the key number has a relative uncertainty at best of 1 part in 3600 and so the answer has a relative uncertainty at least of this magnitude (i.e., about 2.5 parts in 8900). The objective in a calculation is to express the answer to at least the precision of the least certain number, but to recognize the magnitude of its uncertainty. (Similarly, in making a series of measurements, one should strive to make each to about the same degree of relative uncertainty).

If the magnitude of the answer with regard to decimal or sign is smaller than that of the key number, one additional figure may be carried in the answer in order to express the minimum degree of uncertainty, but it is written as a subscript to indicate that it is more doubtful.

Example 6.10 *Give the answer of the following openion to the maximum of significant figures and indicate the key number.*

$$\frac{42 \cdot 68 \times 891}{132.6 \times 0.5247} = 546.57.$$

Solution. The key number is 891. Since the absolute magnitude of the answer is less than the key number, it becomes 546.6 The last 6 is written as a subscript to indicate it is more doubtful. Again, the key number has a relative uncertainty of about 1 part in 900, so the answer has uncertainty of at least 6 parts a 5500 (0.6 parts in 550).

In multiplication and division, the answer from each step of a series of operation can statistically be rounded to the number of significant figures to be retained in the final answer. But for consistency in the final answer, it is convenient to carry one additional figure until the end and then round off.

Logarithms. In changing from logarithms to antilogarithms and *vice vers*, the number being opetated on and the logarithm mantissa have the same number of significant figures. All zeros in the mantissa are significant. Suppose, for example, we wish to calculate the pH of a $2 \cdot 0 \times 10^{-3}$ M solution of HCl from pH = -log $[H^{+}]$. Then, pH= $-\log 2.0 + 10^{-3}$ = - (- 3+0.30) = 2.70. The -3 is the characteristic (from 10^{-3}), a pure number determined by the position of the decimal. The 0.30 is the mantissa for the logarithm of 2.0 and therefore has only two digits. So, even though we know the concentration to two figures, the pH (the logarithm) has three figures. If we wish to take the antilogarithm of a mantissa, the corresponding number will likewise have the same number of digits as the mentissa. The antilogarithm of 0.072 (contains three figures in mentissa .072) is 1.18, and the logarithm of 12.1 is 1.083 (1 is the characteristic, and mantissa has three digits .083).

Rounding off. If the digit following the last significant figure is greater than 5, the number is rounded up to the next higher digit. If it is less than 5, the number is rounded to the present value of the last significant figure :

9.47 = 9.5

9.43 = 9.4

If the last digit is a 5, the number is rounded off to the nearest even digits :

8.65 = 8.6,

.75 = 8.8,

8.55 = 8.6.

REFERENCES

W.J. Youden. *Statistical Methods for Chemists*, New York : John Wiley & Sons, Inc ; 1951.

H.C.Batsons, *An introduction to Statistics in the Medical Sciences*, Minneapolis : Burgess Publishing Co ; 1956.

H.A. Laitinen and W.E.Harris. *Statistics in Quantitative Analysis, Sampling and Chemical Analysis*, 2nd ed. New York : McGraw—Hill,

Inc ; 1975, Chapters 26 and 27. The latter chapter discusses the statistics of sampling.

L.S.Nelson. *Experimental Determination of an Error Distribution*, J.Chem. Educ.33(1956) 126.

R.B. Dean and W.J.Dixon. *Simplified Statistics for Small Number of Observations*, Anal. Chem. : 23 (1951) 636.

W.J. Blaedel, V.W. Meloche, and J.A. Ramsay. *A Comparison of Criteria for the Rejection of Measurements*, J. Chem. Educ. : 28(1951) 643.

A.G.Worthing and J.Geffner. *Treatment of Experimental Data*, New York : John Wiley & Sons, Inc : 1943. Chapter 12, Correlation Coefficient.

PROBLEMS

1. Replicate analyses of a sample of blood meal has produced the following percentage for nitrogen content : 4.16, 4.21, 4.14, 4.12. For these data, calculate the following.
   (*a*) The average value for percent nitrogen.
   (*b*) The median value for percent nitrogen. Ans. 4.17%
   (*c*) The average deviation of these results in absolute terms.
   (*d*) The average deviation of these results in parts per thousand.
   (**Ans.** 7 parts per thousand)

2. Calculate the average deviation for each of the following sets of data in absolute and relative terms.

A	*B*
93.6 percent	10.02 percent
93.7	10.01
93.5	10.09

   (*a*) Which set possesses the larger absolute deviation ? (**Ans.** A)
   (*b*) Which set possesses the larger relative deviation ? (**Ans.** B)

3. How many significant figures are there in the following :
   (*a*) 0.062005, (*b*) 31.4, (*c*)0.00625, (*d*) 2.81, (*e*)0.60025 (*f*) 41.3798, (*g*) 60.025, (*h*) 3.14×10^{-2} (*i*) 4.2 (*j*) 620.1 (*k*) 96.494 (*l*) 44.21 (*m*) 0.0003014 (*n*) 35.458 (*o*) 91.22 (*p*) 0.001 (*q*) 0.014334 (*r*) 1.008 (*s*) 0.002605 (*t*) 6.111 (*u*) 2.6528 (*v*) 0.0341 (*w*) 0.0101 (*x*) 0.01922.

4. Express each of the numbers in problem 3 to the three significant figures.

5. Express the result of each of the following calculations to the proper number of significant figures :

   (*a*) $14.37 \times 6.4 = 92.5428$

   (*b*) $0.0613 \times 0.4044 = 0.0247892$

   (*c*) $0.0613 + 0.4044 = 0.151582$

   (*d*) $0.841 + 297.2 = 0.00282974$

   (*e*) $4.1374 \times 0.841/297.2 = 0.0117077$

6. Express the result of each of the following calculations to the proper number of significant figures :

   (*a*) $4.1374 + 2.81 + 0.0603 = 7.0077$

   (*b*) $4.1374 - 0.0603 = 4.0771$

   (*c*) $4.1374 - 2.81 = 1.3274$

   (*d*) $4.1374 - (2.81 + 0.0603) = 1.2671$

7. Express the result of each of the following calculations to the proper number of significant figures :

   (*a*) $4.178 + 0.0037/60.4 = 0.0692323$ **(Ans. 0.0692)**

   (*b*) $4.178 \times 0.0037/60.4 = 0.000255937$

   (*c*) $4.178 - 4.032/1.217 = 0.119967$

   (*d*) $4.178 + 4.032/1.217 = 6.74609$

   (*e*) $(6.3194 - 1387)(204.2)/0.2148 = 2073.08$

8. Evaluate the probable uncertainty in the result for each of the calculations in problem 6.

9. When a method of analysis was tested upon samples of varying size the following data were obtained :

Sample size, gram	*Weight of component found, gram*
1.0021	0.4171
0.7997	0.3329
0.6014	0.2506
0.4011	0.1674
0.1995	0.0837
0.1008	0.0426

What, if any thing, do these data reveal relative to the existence of determinate error in the analysis ?

For each of the following set of data :

7.031	31.41	63.74	90.91
7.339	30.64	63.62	90.42
7.126	31.52	63.93	90.31
7.027	31.18	63.68	90.24

(*a*) Apply the Q test to see if rejection of the outlying result is justified.
(*b*) Report the best value for each set and define your choice.
(*c*) Calculate the 95% confidence interval for each set.

10. The following replicate molarities were obtained when standardizing a solution : 0.1067, 0.1071, 0.1066, 0.1050, Can one of the results be discarded as due to accidental error ?

11. Replicate samples of a silver alloy are analyzed and determined to contain 95.67, 95.61, 95.71 and 95.60% Ag. Calculate
(*a*) the average deviation in ppt,
(*b*) the standard deviation, and
(*c*) the coefficient of variation (%) of the individual results. Also calculate
(*d*) the standard deviation of the mean.

12. The following molarities were calculated from replicate standardization of a solution ; 0.5026, 0.5129, 0.5023, 0.5031, 0.5025, 0.5032, 0.5027, 0.5026. Assuming no determinate errors, what range are you 95% certain that the true value of the molarity falls within ?

Ans. 0.5025 to 0.5029 M

13. A new method for the determination of cholesterol in serum is being developed in which the rate of depletion of oxygen is measured with an oxygen electrode upon reaction of the cholesterol with oxygen, when catalyzed by the enzyme cholesterol oxidaze. The results for several samples are compared with those of the standard Lieberman colorimetric method. From the following data, determine by the t-test if there is a statistical significance between the two methods.

Sample	*Enzyme method (mg/dl)*	*Colorimetric method (mg/dl)*
1	305	300
2	385	392
3	195	185
4	162	152
5	478	480
6	455	461
7	238	232
8	298	290
9	408	401
10	323	315

14. The following replicate calcium determinations on a blood sample using atomic absorption spectrophotometry (*AAS*) and a new colorimetric method were reported. Is there a significant difference in the precision of the two methods ?

AAS (*mg/dl*)	*Colorimeiric* (*mg/dl*)
10.9	9.2
10.1	10.5
10.6	9.7
11.2	11.5
19.7	11.6
10.0	9.3
	10.1
	11.2
Mean = 10.4	Mean = 10.4

CHAPTER 7

pH and Buffers

Definition of pH and its measurement

pH may be defined according to Sorensen (1909) as the negative logarithm of hydrogen ion concentration expressed in molarity *i.e.*,

$$pH = -\log[H^+] \qquad ...(1)$$

where $[H^+]$ is the concentration of hydrogen ions in moles. It is realized that activity rather than the concentration of an ion determines the e.m.f. of a galvanic cell of the type common employed to measure pH and hence pH may be defined more accurately as

$$pH = -\log aH^+ \qquad ..(2)$$

where aH^+ = activity of hydrogen ions

The measurement of pH by the e.m.f. method gives values corresponding to more closely to the activity than the concentration of hydrogen ions. It can be shown that pC_H (H^+ expressed in concentration terms) value is nearly equals to $-\log 1{\cdot}1aH^+$ and hence

$$pH = pC_H + 0{\cdot}04 \qquad ...(3)$$

The equation is useful practical formula for converting table of pH based on the Sorensen scale to an appropriate activity basis, in line with the practical definition of pH given below :

The modern definition of pH is an operational one and is based on the work of standardization and the recommendations of the US National Bureau of Standards (NBS). The NBS and British definition are consistent in many respects. British definition is confined to one standard solution of potassium hydrogen phthlate whereas NBS definition extends to a number of standard solutions.

The difference in pH between two solutions *A* (a standard) and *B* (an unknown) at the same temperature with the same reference electrode and with hydrogen electrode at the same hydrogen pressure is given by :

$$pH_{(A)} - pH_{(B)} = \frac{E_B - E_A}{2{\cdot}3026\ RT/F} \qquad ...(4)$$

where E_B is the e.m.f. of the cell :

H_2. Pt | solution *B* || 0·5 M KCl | Reference electrode and E_A is the e.m.f. of the cell :

H_2, Pt solution *A* || 3.5 M KCl | Reference electrode.

The pH difference is thus a pure number and the pH scale is a series of numbers which express the degree of acidity (or alkalinity) of a solution.

In British standard *A* is 0.05 M solution of potassium hydrogen phthalate, the pH of which is 4.00 at 15°C and at other temperature '*t*' between 0 to 55°C the pH is given by the equation

$$\text{pH} = 4{\cdot}00 + \frac{1}{2}\left[\frac{t - 15}{00}\right]^2 \qquad ...(5)$$

In *NBS*, potassium hydrogen phthalate is also a standard solution. The standard data for other four standard solutions (*NBS*) is given in the Table 7.1. Concentrations of standards have been expressed in molal basis (moles or solute per kilogram of solution). These standards have been accepted by *IUPAC*. For preparing these solutions, distilled water free from CO_2 (pH = 6.7 to 7.3) and reagents with highest purity, *i.e.*, A.R. products should be used. The preparations of these solutions are also given.

(*i*) *Saturated potassium hydrogen tartrate* ($KHC_4H_4C_6$) *solution.* The pH is insensitive to changes of concentration and the temperature of saturation may vary from 22 to 28 ° C, the excess of solid must be removed. The solution does not keep for more than a few days unless a preservative (crystal of thymol) is added.

(*ii*) *0.05 M Potassium hydrogen phthalate* ($KHC_8H_5O_4$). Dissolve 10.21 g of solid compound (dried below 130 ° C) in distilled water and dilute to 1 kg. The solution should be replaced after 5 to 6 weeks.

(*iii*) *0.025 M Phosphate buffer*. Dissolve 3.4 kg of KH_2PO_4 and 3.55 g of Na_2HPO_4 (dried for two hours at 110—130 ° C) in CO_2 free distilled water and dilute to 1 kg.

(*iv*) *0.01 M Borax*. Dissolve 3.81 kg of sodium tetraborate $Na_2B_4O_7$. 10 H_2O in distilled water and dilute to 1 kg.

Table 7.1. pH of NBS Standard

Tempe-rature °C	$KHC_4H_4O_6$ (*Satd at 25° C*)	$KHC_8H_5O_4$ 0.05 *M*	0.025 *M* KH_2PO_4+ 0.025 *M* Na_2HPO4	0.01 *M* *Borx* ($Na_2B_4O_7$ 1 OH_2O)	$Ca(OH_2)$ *satd at 25° C*	0.05 *M* $KH_3(C_2O_4)8$ $2H_2O$
	Primary standard			*Second standard*		+
1	*2*	*3*	*4*	*5*	*6*	*7*
0	—	4.00	6.98	9.46	13.43	1.67
5	—	4.00	6.95	9.40	13.21	1.67
13	—	4.00	6.92	9.33	13.00	1.67
15	—	4.00	6.90	9.28	12.81	1.67
20	—	4.00	6.88	9.23	12.63	1.68
25	3.56	4.01	6.86	9.18	12.45	1.68
30	3.55	4.02	6.85	9.14	12.30	1.69
35	3.55	4.02	6.84	9.10	12.14	1.69
40	3.55	4.03	6.84	9.07	11.99	1.70
45	3.55	4.05	6.83	9.04	11.84	1.70
50	3.55	4.06	6.83	9.01	11.70	1.71
55	3.55	4.08	6.83	8.99	11.58	1.72
60	3.56	4.09	6.84	8.96	11.45	1.72
70	3.58	4.13	6.85	8.92	—	1.72
80	3.61	4.16	6.86	8.89	—	1.74
90	3.65	4.21	6.88	8.85	—	1.77
95	3.67	4.23	6.89	8.83	—	1.88

(*v*) *Saturated $Ca(OH)_2$ solution.* This is secondary standard solution and prepared by shaking large excess of $Ca(OH)_2$ in distilled water at 25° C. Filter through sintered glass crucible G-3 and store it in polythene bottle. Avoid entering of CO_2 into the solution. If turbidity appears, replace the solution. The solution is 0.023 M at 25°C (pH = 12.45), 0.0211 M at _O°C (pH = 12.47) and 0.0195 M at 30°C (pH = 12.44).

(*vi*) *0.05 M Potassium tetoxalate* ($KHC_2O_4 \cdot H_2C_2O_4\ 2H_2O$). Dissolve 12.70 g of $KHC_2O_4H_2C_2O_4\ 2H_2O$ in water and dilute to 1 kg. Sample must not be dried above 50 ° C.

The pH of the standard solutions is measured with the help of pH meter using glass electrode together with a saturated colomel reference electrode.

Basic principle and components of a pH-meter.

When a metal *M* is dipped into a solution containing its own ion M^{n+} an electrode potential is established, the value of which is given by the Nernst equation :

$$E = E^{\circ} + \frac{RT}{nF} I_n\, aM^{n+} \qquad ...(5)$$

where E° is the standard electrode potential of the metal. *E* is measured by combining the electrode with reference electrode.

There are two methods used in making experimental measurements :

1. **Direct potentiometry.** In this, a single measurement of potential of the cell is made and this is sufficient to determine concentration of the

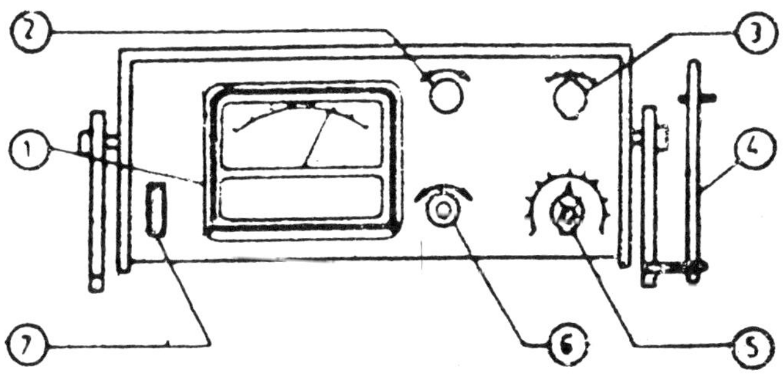

Fig. 7.1. pH-METER

ion. This is used for measurement of pH of aqueous solution. Now a days it is also applied to determine other ions through the use of ion-selective electrodes.

2. **Potentiometric titration.** The ion may be titrated and potential is measured as a function of the volume of the titrant and measurement of potential is used to detect the equivalence point of a titration. It can be used for all types of oxidation-reduction titrations.

The Elico pH-meter model L 1 — 10 T is one of the many models available in the market for routine use. This has been selected to explain the use of the controls of the ph-meters and their use.

Description of Controls.

Front controls, (See Fig. 7.1)

I. Indicating Meter. (1)

It displays meter reading.

II. See Zero. (2)

It is the nature of this type of d.c. amplifiers to show some output reading even without any input signal and also to drift from the zero reading from time to time. After switching ON the instrument, the pointer of the meter is set to mV or pH position by means of this control and also the subsequent drift is corrected by this. The instrument is designed to keep this drift to a minimum, say less than 0.05 pH or 10 mV. This control should be used only when the 'SELECTOR', switch is in 'ZERO' position.

III. Selector. (3)

By means of this switch the function of the instrument is selected, *i.e.* to correct ZERO drift, pH measurement or mV measurement and their different ranges. In its ZERO position it disconnects the electrode pair from the amplifier. In the 0–7 and 7–14 pH position, it brings the asymmetry potential control (SET BUFF) into circuit besides the pH electrodes.

Precaution. 'This should not kept in position other than 'ZERO' when the pH electrode or mV source are not present in solution.

IV. Electrode Sliding Support. (4)

It is provided to place the solution under test in a beaker and it up the electrode on the electrode clip. Electrode clip is spring loaded and can be moved upward and downward to facilitate the depth of immersion of electrode, which should be around 3 cms.

V. Temp. Compensate. (5)

This is a stepless control which compesates for variation in pH when set at the temperature between 0°C and 100°C of the solution.

VI. Set Butter. (6)

This control introduces a potential to cancel the asymmetry potential of the glass electrode. This control is to be operated while calibrating or standardizing the instrument with a standard buffer solution.

VII. On/off Switch. (7)

The instrument is switched ON or OFF by throwing this to proper position.

VIII. Glass Electrode Bush. (8)

Insert the glass electrode jack in the GE both till the touch of the internal spring leaf is felt and firmly secure it by means of thumb screw.

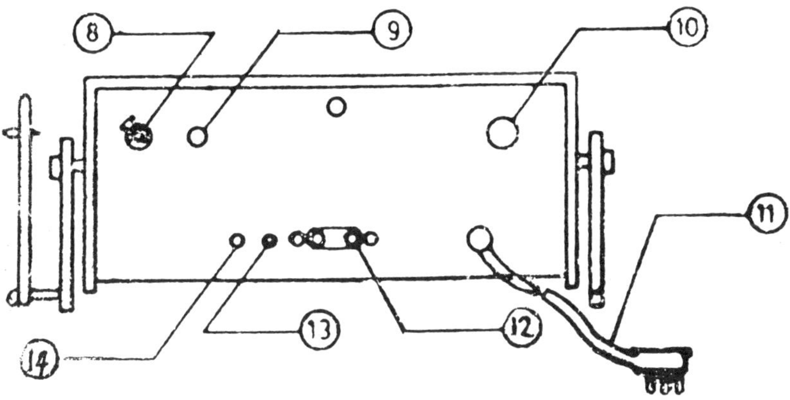

Fig. 7.2. pH-METER

IX. Reference Electrode Terminal. (9)

Connect the reference electrode pin to the terminal.

X. Main Fuse. (10)

This is a protecting device which blow off when the instrument draws excess current under faulty conditions. The fuse catridge can easily be replaced by unscrewing the top cap.

XI. Power Cable. (11)

This is 3-core mains cable terminated in a 3-pin plug. The RED and BLACK leads are the phase and neutral of the line. The Green/White lead is the earth (or Ground) connected one and should be firmly connected to earthed metal such as water pipe where 3-way power socket is not available.

XII. Recorder. (12)

Terminal strip. From these terminals, an output of 10 mV can be obtained corresponding to the full scale deflection of 7 pH units, the polarity depending upon the acidity or alkali range. This singal is usually sufficient to drive recorders with a 10 mV sensitivity.

XIII. Asymmetry Adjustment (See Figs. 7.1 and 7.2). 13

Due to again, the asymmetry potential of the glass electrode may change thereby rendering the 'SET BUFFER' (No. 6) control range insufficient in one direction or the other. As a result it may not be possible to standardize the instrument with a known buffer solution. In such a case, the following procedure may be followed to bring the 'SET BUFFER' control into operation.

(*i*) Set zero is usual.

(*ii*) Take a known buffer solution, keep the 'SELECTOR' in the proper range (0 — 7 or 7 — 17 pH). For example take 4 pH Buffer solution.

(*iii*) Set the 'SET BUFFER' control approximately at its midpoint of rotation.

(*iv*) Then, by means of the 'Asy Adj' preset control, bring the meter to read the pH value of the buffer. This control also provides an (auxiliary) asymmetric potential adjustment range of ± 150 mV or 3 pH.

(*v*) This operation disturbs the 'ZERO' reading. To correct it, put, the 'SELECTOR' in 'ZERO' position and adjust the zero by means of the front SET ZERO control. Now try to standardize with the buffer. The above operations have to be repeated a couple of times.

XIV. 'Zero'.

Due to ageing of the components of the SET ZERO control (No. 2) may become ineffective in bringing the pointer to read zero. In such a case, the ZERO control is useful. The following procedure should be followed to bring the instrument to normal condition.

(*i*) Keep the SELECTOR (No. 3) in ZERO position.

(*ii*) Set the ZERO control on the front panel at its mid-point.

(*iii*) Now adjust the ZERO control by means of a screw driver to bring the pointer to zero.

Preparation and maintenance of electrodes for use :

(*a*) **Fibre junction calomel reference electrode.**

1. Remove the rubber cap and any protective coating at the tip of electrode before use.

2. A rubber sleeve covers the filling hole. Side down this sleeve to keep the hole open during operation. This ensures free flow of electrolytic solution.
3. The electrolytic solution flows out very slowly through the fibre junction. So this solution must be replaced with a fresh saturated potassium chloride solution.
4. If the electrode is allowed to dry out, permanent damage may result.

(*b*) **Glass electrode**

1. Soak the glass bulb in N/10 hydrochloric acid for 24 hours to activate the conductive memberane.
2. Rinse the bulb in distilled water.
3. Shake the filling solution into the bulb.

Precaution.

1. Do not handle the glass memberane or subject it to abrasion of any kind as this may cause sluggishness or span error.
2. See the range of pH specified before use. Permanent damage will happen to electrode if used outside the range.
3. See the usable range of temperature before use. The electrode will lose the response if used beyond.

Maintenance.

1. Clean the Electrode after use by rising it in distilled water. Strongly adhering impurities can be removed by careful wiping the electrode with a moist pad of cotton wool. Grease films can be removed with carbon tetrachloride, benzene or soap solution, after which the membrane should be soaked for 24 hours in distilled water.
2. When not in use, store the electrode in distilled, water, or preferably in a medium similar in pH to the test sample. In the event of prolonged storage, the electrode can remain dry, but it must be reactivated before use by soaking it for 24 hours in N/10 hydrochloric acid.
3. Keep the terminal plug clean and dry at all times, otherwise loss of insulation can be expected resulting in span error, or, in the case of a dead short, on response at all.

BUFFER SOLUTIONS.

A solution which resists any change in its pH value on dilution or an addition of an acid or alkali is called a *buffer solution* and it keeps its pH value fairly constant.

The *buffer capacity* may be defined as the number of moles, x, of acid (base) which when added to one litre of the buffer solution changes its pH by unity, *i.e.*,

$$\text{Buffer capacity} = \frac{dx}{dpH}$$

pH is the negative logarithm of hydrogen ion concentration *i.e.*,

$$pH = -\log [H^+]$$

The constituents of buffer solutions are weak acids or weak bases and their respective salts with strong alkalis or strong acids. When a small quantity of, for example, hydrochloric acid is added to a solution containing appreciable amounts of acetic acid and sodium acetate, H^+ ions from the added acid combine with the acetate ions present in the solution to form almost unionized acetic acid. Thus the effect of added H^+ ions to change the concentration of H^+ ions of the solution is nullified and the pH of the solution remains constant. Again if one adds a small quantity of NaOH, added OH^- ions react with the acetic acid in the solution to produce acetate ions and water, thereby maintaining the H^+ ions concentration at the initial level. It is needless to mention that the similar addition of acid or alkali or the presence of impurities like CO_2 In water to pure salt solution would markedly change the pH of the solution.

In a solution containing a weak acid, the following equilibrium will exist :

$$HA \rightleftharpoons H + A^-$$

and $$[H^+] = K_a \frac{[\text{acid}]}{[\text{anion}]} \text{ By Law of Mass Action)}$$

When the solution of weak acid also contains an appreciable concentration of the salt of the acid, the anion concentration may, be regarded as that of the salt.

∴ For a buffer solution of a weak acid and its salt,

$$[H^+] = K_a \frac{[\text{acid}]}{[\text{salt}]}$$

or $$pH = pK_a + \log \frac{[\text{salt}]}{[\text{acid}]} \quad ...(1)$$

Similarly for a buffer solution composed of a weak base and is salt.

$$pOH = pK_b + \log \frac{[salt]}{[acid]} \quad ...(2)$$

or

$$pH = pK_w - pK_b - \log \frac{[salt]}{[acid]}$$

The preparation of buffer solution is based on equations 1 and 2. The resistance to the change in pH by the addition of any acid or alkali is maximum at [salt] = [acid] or when pH = pK_a (pOH = pK_a). In other words, buffer capacity is maximum when the salt concentration becomes equal to that of the acid concentration in the mixture and falls off on each side. However, satisfactory buffer properties of an acid (or base) and its salt generally extend over a pH (or pOH) range equal to pK + 1 to pK — 1. This in order to prepare buffer solutions covering the complete range of pH values, it would be essential to use acids or bases of varying K_z values. (However, for very low or high pH values concentrated solutions of strong acids or strong bases may be employed to act as buffer solutions.

Preparation of buffer solutions. A buffer solution of any pH can be prepared from equations 1 and 2. For the lower pH values, an acid is chosen such that its pK_a is near to the desired pH. Thus the concentration of the acid and its salt are to taken that the mixed solution gives the desired pH (Eq. 1). Similarly buffer solutions of higher pH values can be obtained by using weak bases of pK_b near to pOH or pK_w – pH. Alternatively, when it is desired to prepare different buffer solutions within a certain pH range it is preferably done by adding different amounts of the acid (base) to a fixed amount of the salt and calculating the pH in each case by equations 1 — 2. This procedure has the advantage in that the ionic strength or salt content is kept constant, when using monobasic acids or monoacidic bases. In the case of polybasic acids or polyacidic base constancy of ionic strength is maintained by addition of calculated amounts of potassium or sodium chloride. The following buffer tables will be found useful for the preparation of buffers of different pH values.

Sodium Acetate-Hydrochloric Acid Buffer. 1.0 M Sodium Acetate ; 10 M Hydrochloric acid. To 20 ml of sodium acetate solution and the following quantities (ml) of hydrochloric acid and distilled water.

pH	*HCl*	H_2O
1.0	29.6	50.4
1.5	23.4	56.6
2.0	21.0	59.0
2.5	20.2	59.8
3.0	19.6	60.4
3.5	18.4	61.6
4.0	15.6	64.4

Acetic Acid-Sodium Acetate Buffer. (0.2 M Acetic acid ; 0.2 M Sodium accetate)

pH	*Acetic acid, ml*	*Sodium acetate, ml*
3.5	94	6
4.0	82	18
4.5	56	44
5.0	30	70
3.2	20	80
5.5	12	88
5.6	10	90

Phosphate Buffer. (*M*/15 Potassium dihydrogen phosphate ; *M*/15 Disodium hydrogen phosphate)

pH	KH_2PO_4, *ml*	Na_2HPO_4, *ml*
5.5	96	4
6.0	98	12
6.4	72	28
5.5	69	31
6.6	63	37
6.8	50	50
7.0	38	62
7.5	15	85
7.9	7	93
8.0	6	94

Borate Buffer. [0.05 M Borax ; 0.2 M Boric acid (0.05 M in sodium chloride]

pH	*Borax, ml*	*Boric acid, ml*
7.0	6.0	94.0
8.0	27.5	72.5
9.0	80.0	20.0
10.0	97.5	2.5
11.0	98.4	1.6

Experiment. To prepare different buffers and measure their pH by a pH-meter, and compare the values with theoretically obtained values.

Procedure. Prepare different buffer solutions as per details in the above buffer tables. Measure their pH values using pH-meter. Calculate the values using equations (1) and (2). Calculate the deviations for each value (positive or negative).

CHAPTER 8

Solvent Extraction

Solvent extraction no doubt is one of the oldest known and separation technique. The first recorded instance of the use of solvent extraction was by E. Peligot in 1842, who for the purification of uranyl nitrate, obtained from pitch blende, extracted it from aqueous solution into ether. In 1872, Berthelot and Jungfleisch reported that "a solute, partially soluble in two immiscible organic solvents, distributes when put in presence of each solvent following some simple relation". Now a days in analytical chemistry solvent extraction is the most wide spread separation technique.

In 1891 W. Nernst extended the observation of Berthelot and coworkers and gave a quantitative relation for the distribution of the solute between two immiscible solvents. The relation is

$$K_D = \frac{Cx_2}{Cx_1} \quad ...(1)$$

where Cx_1 and Cx_2 are the concentrations of solute in liquid 1 and liquid 2, respectively K_D is called the partition co-efficient or the distribution coefficient. Partition coefficient and the distribution efficient differ from each other if the molecular form of the solute is not the same in two immiscible solvents. Let us clarify these terms further.

Distribution coefficient or ratio or constant (D)

$$D = \frac{\text{Total solute concentration in organic phase}}{\text{Total solute concentration in aqueous phase}} \quad ...(2)$$

$$D = \frac{\Sigma\,[M]_{org}}{\Sigma\,[M]_{ag}}$$

$$= \frac{[M_I]_{org} + [M_{II}]_{org} + [M_{III}]_{org} + ...}{[M_I]_{ag} + [M_{II}]_{ag} + [M_{III}]_{ag} + ...}$$

where 'M' is the solute which distributes in organic and aqueous phases and M_I, M_{II}, M_{III}, etc. are the different forms of solute in two phases.

Partition constant or coefficient (K_D)

$$K_D = \frac{[M]_{org}}{[M]_{ag}}$$

$$= \frac{\text{Concentration of 'M' in organic phase}}{\text{Concentration of same form of } M \text{ in water}} \quad ...(3)$$

To distinguish further let us discuss the partition of a weak acid in two immiscible phases.

Consider and acid HR. Let us assume that acid is monomeric in both solvents and anion part R^- is not extracted into organic phase, then

$$K_D = \frac{[HR]_{org}}{[HR]_{ag}} \quad ...(4)$$

while
$$D = \frac{[HR]_{org}}{[HR]_{ag} + [R^-]_{ag}} \quad ...(5)$$

The weak acid dissociates in water as

$$HR_{ag} \rightleftharpoons H^+_{ag} + R^-_{ag}$$

$$K_a = \frac{[H^+]_{ag}\,[R^-]_{ag}}{[HR]} \quad ...(6)$$

where K_a the dissociation constant of acid HR. From Eq. (6)

$$[R^-] = \frac{[HR]^- \cdot K_a}{[H^+]}$$

Putting this value of $[R^-]_{ag}$ in Eq. (5), we have

$$D = \frac{[HR]_{org}}{[HR]_{ag}} \left(\frac{1}{1 + \dfrac{K_a}{[H^+]_{ag}}} \right) \quad ...(7)$$

If $[H^+]_{ag}$ is increased i.e., at lower pH, HR is not dissociated, i.e. K_a approaches to zero then

$$D = K_D$$

Example 8.1. *The K_D for benzoic acid between an organic solvent and water is 1.88. Calculate the value of distribution ratio of benzoic acid in between these solvents at PH 6.0.* ($K_a = 6.32 \times 10^{-5}$)

Solution. The equilibrium involved in water

$$C_8H_5COOH \rightleftharpoons C_6H_5COO^- + H^+$$

$$K_a = \frac{[C_6H_5COO^-][H^+]}{[C_6H_5COOH]} = 6.32 \times 10^{-5}$$

Since $D = K_D \left(\frac{1}{1 + \frac{K_a}{H^+}} \right)$

then $$D = \frac{1}{1 + \frac{6.32 \times 10^{-5}}{10^{-6}}}$$

$$= 1.88 \frac{1}{64.2} = 2.93 \times 10^{-2}$$

Thus dissociation of benzoic acid in water to benzote ion has considerably reduced the value of $K_D = 1.88$ to $D = 0.0293$.

Example 8.2. *The D for ruthenic acid is found to be constant as 58.4 for the partition of this acid from neutral and acidic solutions into carbon tetrachloride. In 1.0 × 10^{-2} M NaOH the value of D for ruthenic acid is reduced to 7.90. Calculate the value of dissociation constant of ruthenic acid.*

Solution. Equilibrium for ruthenic acid in water

$$H_2RuO_5 \overset{K_a}{\rightleftharpoons} H^+ + HRuO^{-5}$$

$$K_a = \frac{[HRuO^-{}_5][H^+]}{[H_2Ru_5O]}$$

But $$K_D = \frac{[H_2RuO_5]_{org}}{[H_2RuO_5]_{ag}}$$

and $$D = \frac{[H_2RuO_5]_{org}}{[H_2RuO_5]_{ag} + [HRuO_5]}$$

The constancy of $K_D = 58.4$ in neutral and acidic medium confirms that H_2RuO_5 is not dissociated under these conditions.

$$58.4 = \frac{[H_2RuO_5]_{org}}{[H_2RuO_5]_{sg}}$$

Put the values of $KD = 58.4$

$$D = 7.90$$

and of $$[H^+] = \frac{Kw}{[OH^-]} = \frac{10^{-14}}{1 \times 10^{-2}} = 10^{-12}$$

in Eq.(7) $$7.90 = 58.4 \left(\frac{1}{1 + \frac{K_a}{1 \times 10^{-2}}} \right)$$

$\therefore$ $$K_a = 6.40 \times 10^{-12}$$

Separation by solvent extraction is performed when analytical determinations cannot be performed in initial environment and removal to new environmental is desired, if an analysis is to be feasible. Solute is put in presence of two immiscible liquids and is shaken. The two liquids may be brought in contact with one another discretely or continuously, giving rise to three common ways of carrying out an extraction. These are :

(*i*) Batch Extraction
(*ii*) Continuous Extraction
(*iii*) Counter Current Extraction.

The way selected for any particular extraction will depend upon the relative values of D of the components in the original mixture, the equipment available, and convenience.

Batch Extraction. In batch extraction, a liquid (usually water) containing the dissolved solute sample is shaken with another liquid (usually organic) in a closed container (separatory funnel) until equilibrium has been established. The two immiscible phases are allowed to settle and separated mechanically. When the value of D for the desired component is greater than 10 and considerably different from other components of the mixture, a batch extraction is preferred.

The fraction E of the total amount of solute in the system extracted into the organic phase in a single equilibrium stage is

$$E = \frac{W_{org}}{W_{org} + W_{ag}} \quad ...(8)$$

when the extracted species can exist in only one form then

$$D = K_D = \frac{C_{org}}{G_{ag}} = \frac{W_{org}/V_{org}}{W_{ag}/V_{ag}} \quad ...(9)$$

where $W_{org,}$ W_{ag} are the amounts of solute (either weight or moles) in organic and aqueous phases respectively and V_{org} and V_{ag} are the volumes of organic and aqueous phases respectively. C_{org} and C_{ag} represent the total concentrations of solute in organic phase and water respectively.

D can be written as $\dfrac{W_{org}}{W_{ag}} \times \dfrac{V_{ag}}{V_{org}}$ ...(10)

Equation (8) can be Rearranged as

$$E = \frac{1}{1 + \dfrac{W_{ag}}{W_{org}}} \qquad \text{...(11)}$$

Putting the value of W_{org}/W_{ag} from Eq. (10) into Eq. (11),

$$E = \frac{1}{1 + \dfrac{V_{ag}}{DV_{org}}}$$

$$E = \frac{DV_{org}}{DV_{org} + V_{ag}}$$

or

$$\%E = \frac{100\, DV_{org}}{DV_{org} + V_{ag}} \qquad \text{...(12)}$$

If W is the total amount of solute initially present in water, then

$$D = \frac{W - W_{ag}/V_{org}}{W_{ag}/V_{ag}} = \frac{W - W_{ag}}{V_{org}} \cdot \frac{V_{ag}}{W_{ag}}$$

$$DV_{org}\, W_{ag} = (W - W_{ag})V_{ag}$$

$$W_{ag} = W\left[\frac{V_{ag}}{DV_{org} + V_{ag}}\right] \qquad \text{...(13)}$$

Equation (13) tells us the amount of solute non entracted or remained after one extraction.

Similarly

$$W_{org} = W\left[\frac{DV_{org}}{DV_{org} + V_{ag}}\right] \qquad \text{...(14)}$$

can be derived simply by rearranging the above equation for D. Equation (14) gives the amount of solute being extracted in organic phase after first extraction.

Let us now see what happens if aqueous solution is extracted second time with organic solvent and let w'_{ag} and w'_{org} be the amounts of solute remained in water and extracted into organic phase respectively, then

$$D = \frac{w'_{org}/V_{org}}{w'_{ag}/V_{ag}} = \frac{w_{ag} - w'_{ag}/V_{org}}{w'_{ag}/V_{ag}}$$

or
$$D = \frac{\dfrac{W[V_{ag}/DV_{org} + V_{ag}] - w'_{ag}}{V_{org}}}{w'_{ag}/V_{ag}} \qquad ...(15)$$

On rearranging equation (15)

$$w'_{ag} = W\left(\frac{V_{ag}}{DV_{org} + V_{ag}}\right)^2 \qquad ...(16)$$

Similarly it can be shown that

$$w'_{org} = W\left[\frac{DV_{org} \cdot V_{ag}}{(DV_{org} + V_{ag})^2}\right] \qquad ...(17)$$

For analytical purposes equation (17) is not of much use. What is of importance is the combined amount of solute extracted during both first and second eqilibria. Adding equations (14) and (17).

$$w_{org} + w'_{org} = \text{Total amount extracted}$$

$$= \frac{WDV_{org}}{DV_{org} + V_{ag}}\left[1 + \frac{V_{ag}}{DV_{org} + V_{ag}}\right] \qquad ...(18)$$

Thus the equation for the total amount being extracted after '*n*' extraction will be more difficult. However equation for the amount of solute being left after *n* extraction, can be written after generalizing the equation (16) as

$$(w_{ag})_{nth} = W\left[\frac{V_{ag}}{DV_{org} + V_{ag}}\right]^n \qquad ...(19)$$

For the derivation of equations (16), (17), (18) and (19), it has been assumed that D remains constant and does not vary with solute concentration.

Since extraction is an equilibrium process, each contact with a fresh organic liquid removes a constant reaction of the total concentration of the solute. In equation (19), the volume ratio term is exponential in '*n*', from which it follows that for a given volume of organic extractant, V_{org}, more solute is extracted when extracting the aqueous phase with *n* times with V_{org}/n portions of the organic, then when extracted once with entire volume, V_{org}. For equal volume of two phases, there is no linear relationship between the amount of solute being extracted and D. For equal volume of two phases equation (12) can be written as

$$D = \frac{\%E}{100 - \%E} \qquad ...(20)$$

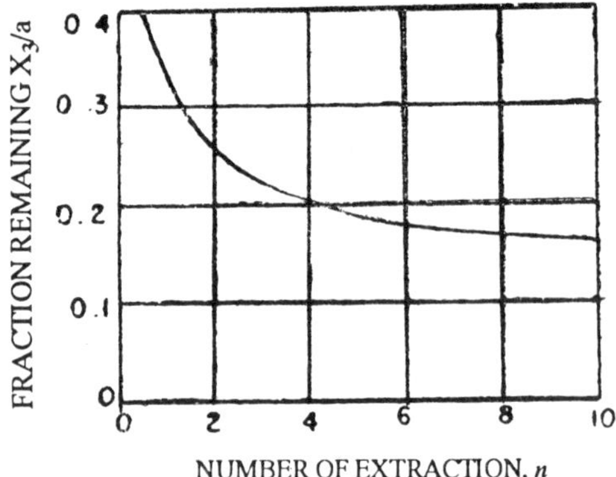

Fig. 8.1. Plot of equation (19), assuming $D = 2.0$ and $V_{ag} = V_{org} = 100$ ml

Example 8.3 *The value of D for $PdCl_2$ between 3 M HCl and TBP is 2.3. How many times must 15.0 ml of a 5.00×10^{-3} M solution of $PdCl_2$ in 3M HCl be extracted with fresh 5.0 ml portions of TBP in order to remove 99.5% of Pd(II) from aqueous phase.*

Solution. Using equation (19) and substituting the various values

$$\frac{w_{ag}}{W} = 0.095 = \left[\frac{15.0}{2.3 \times 5.0 + 15.0}\right]^n$$

$$n = 11$$

Example 8.4. *The D for I_2 between CCl_4 and water is 85. If 100 ml of 1×10^{-3} M of I_2 is extracted twice with 50 ml portions of CCl_4, what mole amount of I_2 will be left in aqueous solution ?*

Solution. Using equation (19) and substituting the values into it.

$$(W)\, I_{2(ag)} = 1 \times 10^{-3} \left(\frac{100}{85 \times 50 + 100}\right)^2$$

Example 8.5. *Calculate the amount of I_2 (m mole) in previous example, that will remain in aqueous, if the solution is extracted with single 100 ml of CCl_4 portion rather than two 50 ml portions.*

$$WI_{2(ag)} = 1 \times 10^{-3}\left(\frac{100}{85 \times 100 + 100}\right)$$

$$= 1.16 \times 10^{-3} \text{ m mole}$$

Solutions of examples 4 and 5 show that single extraction with 100 ml of CCl_4 leaves more I_2 in aqueous than that extracted with two 50 ml portions of CCl_4.

If two solutes *A* and *B* have different values for *D* i.e., D_A and D_B, then they can be separated from one another, regardless of how small the differences in *D* may be. The case and completeness of separation, however, will depend upon the magnitude of the differences in *D* values. For such solute the separation coefficient ' β ' is given by

$$\beta = \frac{D_A}{D_B}$$

and β should be 10^4 or 10^6 for complete separation.

Continuous Extraction. When value of *D* is comparatively small ($D<1$) and considerably different, i.e., $\beta \geq 10^4$ from those of the other components in the mixture, continuous extraction should be the method of choice. In continuous extraction, the extracting liquid (lighter of heavier) is passed continuously through or over the stationary liquid containing the solute. The two phases are then separated physically and one containing unextracted desired solute is again treated for continuous extraction. In continuous extraction the solvent is continuously and rapidly passed and thus equilibrium is never attained. Due to continuous extraction for a longer time most of the solute is ultimately extracted.

The technique and apparatus for continuous extraction is complex and its construction depends upon whether the solution being extracted is heavier or lighter than the extracting liquid and whether the extracting solvent is recyled or not. The various types of continuous extractors employed are shown in following figures.

Extraction with lighter solvent

The extracting solvent is placed in round bottom flask, (*A*) and heated electrically. Its vapours pass into the top of central tube (*B*). They are condensed in condenser (*C*). The drops of condensate fall into long stemmed funnel *D* which goes to the bottom of solution being extracted *E*. Being lighter in nature, the solvent droplets rise up while they come in contact with the solution *E*. Thus the level of solvent in *B* rises and through the side arm passes into the round bottom flask again. Thus the solvent can be recycled.

Extracting with heavier solvents. Fig. 8.3 shows the type of apparatus employed for such extraction. Extracting solvent is placed in round bottom flask (*A*) and heated electrically. Its evapours pass into upper part off the central tube (*B*) and then to condenser (*C*), where they are condensed and droplets of condensate fall into the solution (*D*) which is extracted. Being heavier in nature, they move down while they come in contact with the solution *D* and thus extracting the solute. Hydrostatic pressure will force the extractant through the side–arm (*E*) into the distilling flask (*A*). The process can be repeated until all the solute is extracted.

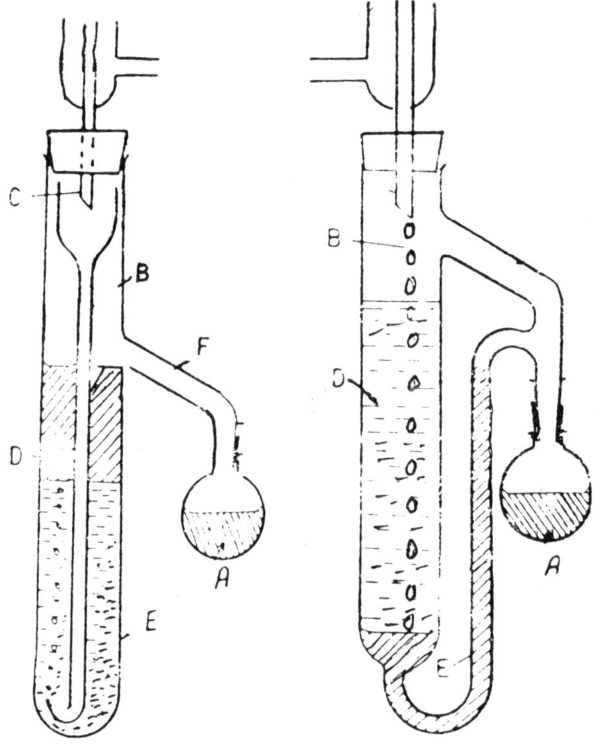

Fig. 8.2. Extractor for use with solvents lighter than water.

Fig. 8.3. Extractor for solvents heavier than water.

Extraction of solid by solvent. When a solid is to be extracted with a solvent, a Soxhlet extractor as shown in Fig. 8.4 or its modification is employed. The solid which is to be extracted is powdered and placed in a filter paper or cloth or paper thimble (*E*) in the central chamber (*A*). The extracting solvent is placed in flat bottom flask (*B*) and electrically heated. Its vapours rise through the side arm (*C*) into the condenser (*D*), where they are condensed and condensate fall on the solid sample in (*E*). The solvent remains into (*E*) and level of it continuously rises and extracts the solute and the level rises to the top of side arm (*F*). At this time, the hydrostatic pressure inside the main chamber causes the solvent to be siphoned back to distilling flask (*B*). This process is continued till all the solute is extracted.

Continuous extraction usually extends for a longer time and efficiency of extraction *depends upon the contact time between two solvents.* Thus the solution properties that will effect the contact time are : (*i*) viscosities of two phases, (*ii*) interficial tension, (*iii*) the relative densities of two solvent, (*iv*) the size of the area of contact between two solvent. (*v*) the rate of attainment of equilibrium, (*vi*) the relative volumes of two solvents and (*vii*) the value of distribution ratio.

Since continuous extractor works for a long time, solutes with quite low *D* values can be extracted.

Counter Current Extraction. In such type of extraction the solvent and the liquid which is being extracted move continuously in opposite direction and thus the solute distributes in the two oppositely moving phases. When the extraction is complete the phases are separated physically. The liquid containing the unextracted solute is used in subsequent operation. Equilibrium is not attained in this technique. This is the most efficient technique for separating mixture containing solutes with *D* values very similar to one another.

Continuous extraction is usually carried out in the apparatus as shown in Fig. 8.5.

The liquid containing the sample (mash) is placed in the flask (*A*) and circulated using circulating pump through column (*B*). The extracting liquid (lighter, ether) is circulated continuously in opposite direction, by heating the solvent in flask (*C*) on a hot plats as shown in the figure. When the lighter liquid rise to the top of the column (*B*) it is siphoned off into distilling flask (*C*). At the end the solute extracted will be in the distilling flask (*C*).

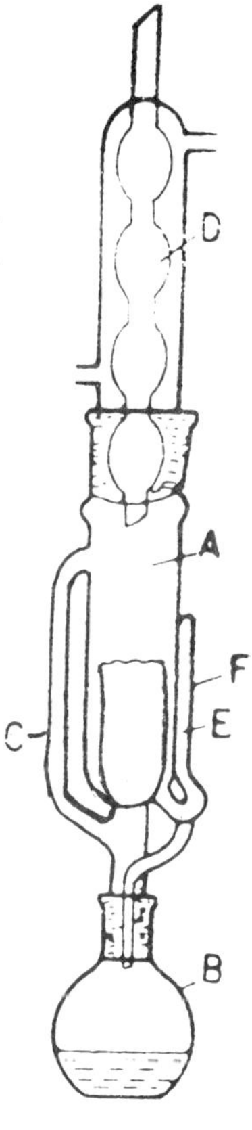

Fig. 8.4

L. Craig and coworker have developed discontinuous counter current distribution technique for counter current distribution. In this technique the two liquid phases are brought in contact with one another in a series of separate vessels in a specially constructed apparatus, shaken, the phases separated and the phases moved to the next vessel. The way in when the apparatus is operated causes in effect the movement of both phases in the opposite direction to one another. At discrete intervals the extraction is stopped to attain the equilibrium. For discussing the Craig's apparatus, let us explain the principle of Craig's discontinuous counter current extraction by taking some simple apparatus.

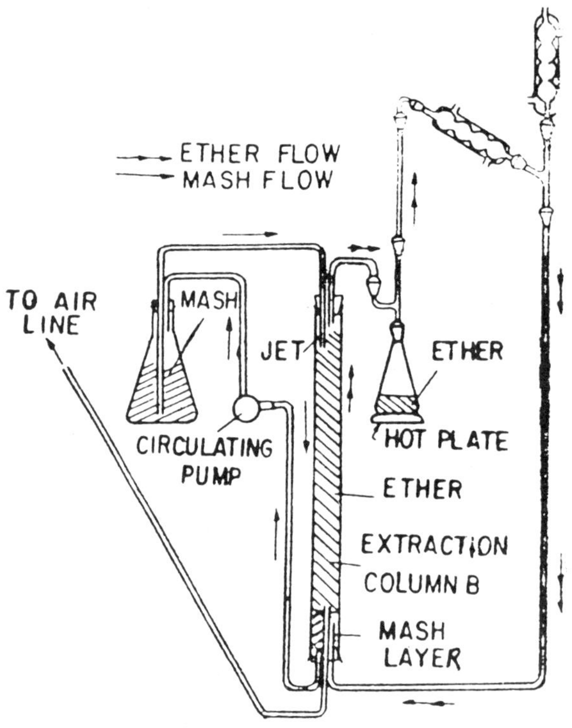

Fig. 8.5.

Let us take a series of 11 separatory funnels numbered from 0 to 10. The funnel numbered to 0 contains 10 ml of the aqueous mixture which is to separated by extraction and the separating funnels 1 to 10 each contain 10 ml of water. To funnel with number $n = 0$ add 10 ml of lighter immiscible organic solvent say diethylether. Shake the funnel until the equilibrium is established, allow the two phase to separate. Transfer the upper organic layer to funnel numbered 1 and add 10 ml of fresh organic layer to funnel 0. Shake the two funnels and allow them to settle. Now transfer upper layer of 1 to funnel 2 and that of 0 funnel to 1. Add 10 ml of fresh diethyl ether to funnel 0 and again shake the funnels numbered 0, 1 and 2 until equilibrium is reached. Allow the layers to separate, transfer the upper organic layer of funnel 2 to funnel 3, of 1 to 2 and that of 0 to 1. Repeat this till the all funnels are used, i.e., you have to carry out the transfer 10 times. The distribution off solute after ten transfer ($n = 10$) will be as shown in Fig. 8.6.

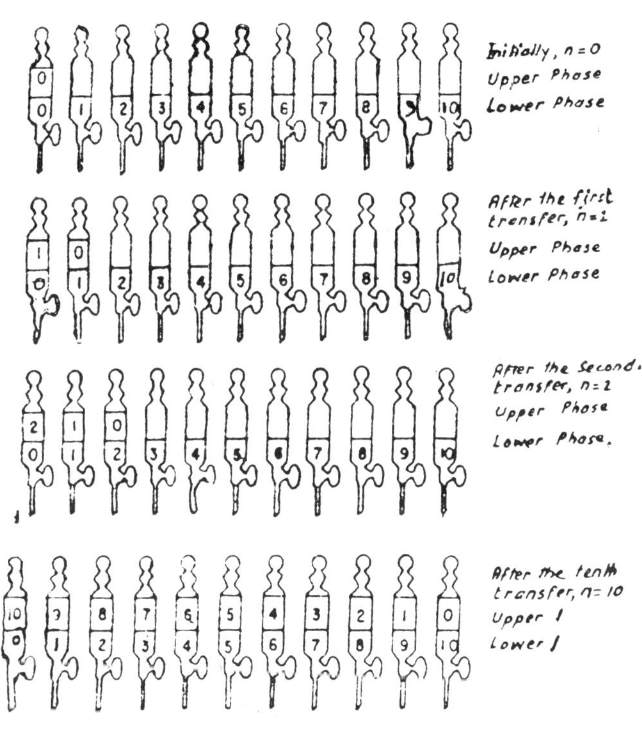

Fig. 8.6.

If the distribution coefficient $D = 1$, the solute will be concentrated in the separatory funnel, if $D>1$, i.e., solute is more soluble in organic phase the solute will be distributed more towards the right and if $D<1$ it will be distributed more towards the left. Usually, the use of series of funnels and pipettes for the liquid transfer is very cumbersome, especially when working with 100.1000 transfers. L. Craig and O. Post had designed an all glass inter connected array of tubes to carry out discontinuous counter current distribution more efficiently.

Choice of Solvents. In solvent extraction the most important question is the choice of solvent. The factors that must be considered in selecting a solvent for extraction are :

1. The solute should be highly soluble in the solvent so the value of D is large. Thus number of extractions required are less.
2. Solvent should be immiscible with water. The idea of immiscibility can be had from the dipole moment (*DPM*) of the solvent. The DPM of the solvent should be quite different from that of water (μ = 1·84 debyes). Solvents such as benzene, hexane, and carbon tetrachloride with μ = 0, have very low water solubilities (~ 0.1 g/100 g), whereas solvent such as diethyl ether (μ = 1·2 debyes), ethyl acetate (μ = 1·8 debyes) or *n*-butyl alcohol (μ = 1·7 debyes) have relatively high aqueous solubilities (6·90, 7·94, 7·80 g/100 g respectively). There is exception, for example chlorobenzeue (μ = 1·6 debyes) is much soluble than benzene.
3. The solvent should have great density difference than that of water, so that separation of layers is clear cut. Carbon tetrachloride (density = 1·59) is much superior than tributylphosphate (*TBP*) (density = 0·98).
4. Viscosity of solvents should be low so that there can be good contact between tow phases when shaking is carried out.
5. The solvent should have sufficiently high boiling point, so that evaporation of the solvent does not cause any difficulty. This is the reason why disopropyl ether is preferred to diethyl ether in extracting H^+ $FeC1_4$ on pair.
6. The solvent should not form stable emulsion wither water.
7. The solute should easily recoverable from the solvent for subsequent operations.
8. Solvent should be cheap and readily available in sufficient purity.
9. Solvent should not be hazardous to health.

A mixture of solvents can also by employed for extraction of certain typical solutes.

Classification of Extraction Systems. In order for a solute to pass from one phase to another phase there must be a net decrease in the energy of the solute. The solute is highly hydrated in water, the hydration energy must be overcome in order to transfer the solute from aqueous phase to organic phase. This hydration energy can be overcome by chemical bond formation, ion pair formation, organic phase solvation or by some combination of these. Depending upon the way by which hydration energy is overcome establishes a general way of classification of extraction systems. The different types of extraction systems are discussed briefly below :

1. Extraction of Non-solvated Solutes. Non-solvated solutes are generally the atoms or covalent molecules that do not have large dipoles so that there is very little interaction in either of the solvent. Such solutes are $HgCl_2$, HgI_2, SbI_3, $AsCl_3$, $GeCl_4$ and $InCl_3$. Separation of such solutes from their mixtures is difficult as the value of D for them is very similar. For example partition coefficients of the two inert gases between water and nitromethane are : He—5·5 and Ne—6·0. These types of solutes however can be removed from other solutes.

2. Extraction of Metal Chelates. One widely used technique for the separation of metal ion is to extract them as their metal chelates. A solution of a chelating agent, HX in an organic solvent is shaken with an aqueous solution of metal ion that will form chelates with it. The equilibria involved are adjusted so that only the desired metal is removed to the organic phase and other metals are not partitioned. A diagram of the pertinent equilibria is represented as :

$$\begin{array}{ccccc} & HX & & MX_n & \\ K_{D_{HX}} & \downarrow\uparrow & \text{———————} & \downarrow\uparrow\ K_{D_{MX}} \quad \text{—————} & \dfrac{\text{Organic}}{\text{water}} \\ & HX \leftrightarrows H^+ + X^- & & MX_n \leftrightarrows M^{n+} + nX^- & \\ & K_a & & K_{st} & \end{array}$$

The distribution of metal refers to the total metal concentration in two phases i.e.,

$$D = \frac{[MX_n]_{org}}{[MX_n]_{ag} + [M^{n+}]_{ag}} \qquad ...(21)$$

Since both chelating agent and chelate are partitioned, there are two partition coefficients involved in the equilibria. The two partition coefficients are ;

$$K_{D_{HX}} = \frac{[HX]_{org}}{[HX]_{ag}} \qquad ...(22)$$

$$K_{D_{MX}} = \frac{[MX]_{org}}{[MX]_{ag}} \qquad ...(23)$$

K_{st} (stability for the formation of complex)

$$= \frac{[MX_n]}{[M^{n+}]\,[X^-]^n} \qquad ...(24)$$

Let us now derive a relation between distribution ratio, partition coefficients, stability constants and dissociation constant.

Get the value of M^{n+} from equation (24) and substitute in equation (21), we have

$$D = \frac{[MX_n]_{org}}{[MX_n]_{ag} + [MX_n]_{ag}/K_{st}\,[X^-]^n} \qquad ...(25)$$

The equlibrium equation

$$HX \underset{}{\overset{Ka}{\rightleftharpoons}} H^+ + X^-$$

$$K_a = \frac{[H^+]\,[X^-]}{[HX]} \qquad ...(26)$$

is for the dissociation of chelating agent.

Get the value of X^- from (26) and substitute in equation (25)

$$D = \frac{[MX_n]_{org}}{[MX_n]_{ag} + [MX_n]_{ag}/K_{st.}\,[K_1\,[HX]/[H^+]]^n} \qquad ...(27)$$

Simplification of (27) does not achieve much. Let us make some reasonable assumptions :

1. Metal chelates are quite stable and most of the metal in an aqueous phase is present in form of metal chelate only. $[MX_n]_{ag}$ is small as compared to $[M^{n+}]_{ag}$. So the first term in the denominator of Eq. (27) is negligible as compared to second term in the denominator, which represents the concentration of uncomplexed metal ion in aqueous phase, i.e., $[MX_n]_{ag} < [M^{n+}]_{ag}$. Therefore Eq. (27) reduces to

$$D = K_{D_{MX}} \cdot K_{st} \cdot (K_a)^n \,\frac{[HX]^n{}_{H_2O}}{[H^+]^n{}_{H_2O}} \qquad ...(28)$$

Concentration of $[HX]_{H_2O}$ at equilibrium is not known. Since we started with concentration $[HX]$ in organic phase it will easier to assume to use the equilibrium concentration of chelating agent in the organic phase. This can be achieved by substituting the partition coefficient relationship for chelating agent into Eq. 28.

$$D = \frac{K_{D_{MX}} K_{st} \cdot K^n{}_a \left[\dfrac{[HX]_{org}}{[K_{D_{HX}}]} \right]^n}{[H^+]^n}$$

$$= K_{D_{MX}} \cdot K_{st} \,\frac{[K_a]^n}{[K_{D_{HX}}]^n} \cdot \frac{[HX]^n_{org}}{[H^+]^n} \qquad ...(29)$$

The distribution ratio is dependent upon :

(*i*) Concentration of chelating agent.

(*ii*) The pH of the system.

(*iii*) The extent to which chelating agent is available to react with the metal ion.

Factors affecting the extraction of metal chelates in organic phases. The various factors which affect the extraction of metal chelate in organic phase are discussed in very brief below :

(*i*) *Concentration of metal ions.* In Eq. (29) concentration of M^{n+} does not appear, so long as the $[M^{n+}]$ lies below the solubility product of MX_n in water, below the solubility of MX_n in organic phase and $[HX]$ is sufficient to complex all the metal ions, the actual concentration of metal ions, does not have any effect on percent extraction.

(*ii*) *Nature of metal extracted.* The value of both $K_{D_{MX}}$ and K_{st} are determined by the forces between M^{n+} and solvent molecules and chelating agents. Hence both these constants will be different for different metal ions. Thus for two metal ions M_1 and M_2 the ratio of their distribution coefficients is given by (see Eq. 29).

$$\beta = \frac{D_1}{D_2} = \frac{K_{DM_1}X \cdot K_{st_1}}{K_{dM_2X} \cdot K_{st_2}} \qquad ...(30)$$

where D_1 and D_2 are the distribution coefficients of metal chelates of M_1 and M_2 with chelating agent HX, under similar conditions.

(*iii*) *Nature of chelating agent.* Since the D depends upon $K_{D_{HX}}$, K_a and K_{st} (see Eq. 29), therefore, substitution of HX chelating agent by another chelating agent will highly affect the distribution coefficient. Table 8.1 gives some of these constants for various chelating agents.

Table 8.1. Constants for various chelating agents of similar structure

Chelating agent	*pKa*	$K_{D_{HX}}$	$K_{D_{MX}}$	K_{st}
Oxine	3.76, 11.19	433	3.02×10^3	25.50
2-Methyl oxine	5.61, 10.16	1670	2.82×10^5	22.82
4-Methyl oxine	4.67, 11.62	1860	3.63×10^3	26.96

The change in value of D cannot be predicted from the above date.

(iv) Effect of the pH of aqueous phase. Since pH affects the dissociation of HX to a greater extent and hence the amount of $[X^-]$ in water to form chelate with metal ion. Higher pH will favour the dissociation of *HX*, chelate formation and extraction. The pH dependent, non-extractable species such as hydroxy species invalidates the above generalization. The magnitude of the change in percent extraction with pH depends upon the formula of the metal chelate, since in Eq. (29) the $[H^+]$ is raised to the power '*n*' for a specific metal-chelating agent system. With a constant organic phase concentration of chelating agent Eq. (29) can be reduced to

$$D = K' [H^+]^{n-} \qquad ...(31)$$

where $$K' = \frac{K_{D_{MX}} \cdot K_{st} \cdot K^{n}_{a} [HX]^{n}_{org}}{K^{n} D_{HX}}$$

Taking log of Eq. (31)

$$\log D = \log K' - n \log H^+ \qquad ...(31\ a)$$
$$= \log K' + n\ pH$$

when $D = 1$ i.e. Concentration of metal in organic and aqueous phase is same, i.e., 50% of the metal is extracted, $\log D = 0$, the pH of solution at this point is called pH one half $(pH_1/2)$ and is a constant characteristic of the system under study.

$$ph_{1/2} = -\frac{\log K'}{n} \qquad ...(32)$$

(*v*) *Concentration of chelating agent in organic phase.* The amount of chelating agent that can be dissolved in the organic liquid is limited by its solubility. Up to the solubility limit the value of *D* will increase with increase in the concentration of chelating agent at constant aqueous phase pH.

(*vi*) *Solubility of chelate in organic phase.* More soluble the metal chelate in the organic phase, the larger is the value $K_{D_{MX}}$ other factors remaining constant during this change. Thus the value of D and thus the percent extraction will increase with increasing solubility of metal chelate in organic phase. Other factors, however, will not remain constant.

(*vii*) *Solubility of the helating agent in water.* Monsolube the chelating agent in the aqueous phase more complexation will be there and hence more extraction of metal chelate in the organic phase will be there. The value of D will increase since $K_{D_{MX}}$ will be lowered see Eq. 22).

(*viii*) *Acid strength of the chelating agent.* As the acid strength of chelating agent increases (i.e., K_a increases), the value of *D* and hence the percent metal extracted will increase. Since the relation between D and K_a is exponential (Eq. 29) acid strength can play a vital role in selection of a chelating agent to use for a particular metal ion extraction. The chelon (chelating agent) with high K_a can be used in more acidic medium (Table 8.1) shows that 2-methyl oxine has larger 'K_a' than 4-methyl oxine and hence for 2-methyl oxine the value of D will be more.

(*ix*) *Stability of metal chelate.* Larger the K_{st} more will be the value of D at constant pH, i.e., we can say that more the stable chelate is, lower will be the pH for certain percentage extraction of metal ion.

(*x*) *Temperature.* Although temperature is not a factor included in Eq. (29). yet it affects the percent extraction of metal ions because all the constants in the expression are temperature dependent. The combined effect of their dependencies gives the temperature dependency of D.

(*xi*) *The effect of the presence of masking agent.* When a masking agent such as cyanide, ammonia, tartrate, NTA or EDTA is present in aqueous phase, the chelon will have to compete with the metal, the value of D will be then lowered (since availability of free metal for reaction with chelon will be less). Thus at the same pH and chelon concentration, the presence of masking agent will decrease the value of D. In presence of masking agent even the pH for extraction will be shifted to higher value.

Separation of metal ions by extraction as their metal chelates. When a solution contain more than one metal ion in the solution, metal ions can be selectively extracted if there is large difference in their extraction constants or their $pH_{1/2}$ values. The separation of two metal ions is defined as being achieved when separation of two metal ions is defined as being achieved when 99% of the extracted one is obtained with only 1% of the unextracted one present as an impurity. For a single batch extraction, in order to achieve this much of extraction, the difference in $pH_{1/2}$ value between two metal ions must be 4 for $n = 1$ (1 : 1 complex) and when $n = 2$ (1 : 2 complex) difference of only 2 pH units and for $n =$ 3 (1 : 3 complex) pH of 1.3 units will be suffice. Thus from the $pH_{1/2}$ values of various metals under similar condition it is possible to predict the separation of metal ion by extraction method. From Eq. (29) for two metals under identical experimental condition of pH and chelon concentration and with identical in n values it follows that

$$\beta = \frac{D_1}{D_2} = \frac{K_{st_1}}{K_{st_2}} \cdot \frac{K_{D_{M_1X}}}{K_{D_{M_2X}}}$$

Table 8.2 gives the pH range for the complete precipitation of various metal ions as their oxinates i.e. the pH for maximum entraction.

Table . 8.2. pH range for formation of oxinate complexes

Metal ion	*Initial precipitation*	*Complete precipitation*
Al^{3+}	2·9	4·7 —9·8
Bi^{3+}	3·7	5·2 —9·4
Cd^{2+}	4·5	5·5 —13·2
Ca^{2+}	6·8	9·2 —13·2
Co^{2+}	3·6	4·9 —11·6
Cu^{2+}	3·0	3·3
Fe^{2+}	2·5	4·1 —11·2
Pb^{2+}	4·8	8·4 —12·3
Mg^{2+}	7·0	8·7
Mn^{2+}	4·3	5·2 —9·5
Ni^{2+}	3·5	4·6 —10·0
Zn^{2+}	3·3	4·4
UO_2^{2+}	3·7	4·9 —9·3

Example 8.6. *The stability constants for copper oxinate and lead oxinate are 5·9 × 10*^{1} *and 9·1 × 10*$^{-9}$ *respectively. Can a solution containing both these ions be separated by a single bath extraction with 0·1 M oxine in chloroform ?*

Solution.
$$\beta = \frac{D_{C_u}}{D_{P_b}} = \frac{K_{C_u}}{K_{P_b}} = \frac{5 \cdot 9 \times 10^1}{9 \cdot 1 \times 10^{-9}} = 6 \cdot 5 \times 10^9$$

Thus there is possibility for their separation from the mixture. However, conditions of pH, etc., are also important.

When the compositions of two metal chelates are same, i.e., *n* is same, then I D_2 is independent of pH. However, when value of *n* is different for different metals, change in pH can be very effective in their separation. For effective separation, the masking agents are used to change the value of *D* of the metal.

3. Extraction of Ion Pairs. A common feature to all the ion-pair extractions is that the solute transfers from concentrated solutions of reagents into polar or oxygenated organic solvents such as alcohols, ethers, ketones, etc. Ion-pair can be divided into the categories :

(*i*) *Simple : simple ion pair.* The category includes the extraction into nitrobenzene of caesium or rubidium tetraphenylboride of metal Fe(III), Be(II), Ga(III), In(III) with benzoate or high molecular weight carboxylic acids into ethyl acetate.

(*ii*) *Simple : simple coordinatior ion-pair formation.* The substance to be extracted is turned in complex ion which is then extracted as one of the ions in an ion-pair. Example include the extraction of $H^+ FeCl_4^-$ in ether and thiocyanate complexes of Cu(II), Co(II), W(VI), Mo(VI) in ether.

(*iii*) *Simple : Chelate ion-pair formation.* In this category a chelate is formed and is then extracted as one of the ion of the ion-pair. Example includes the rhodamine complex of Sb(V) of Ga(III) into ethyl or hexyl alcohols.

Behavour of such systems cannot be mathematically represented. There are numerous reasons for this. The factors that affect the extraction of ion pairs can be summarized as : (i) the effect of acidity, (ii) nature of acid present, (iii) nature of ion-pair extracted, (iv) nature of solvent, (v) concentration of metal ion and (vi) presence of salts in the systems.

4. Extraction of Solvated System. The P—O bond is a highly polar one and is largely responsible for the solvating ability of the organophosphorus compounds. Some of the organophosphorus are :

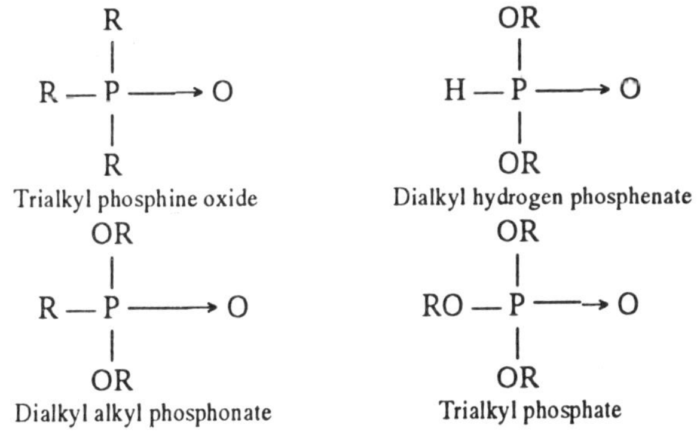

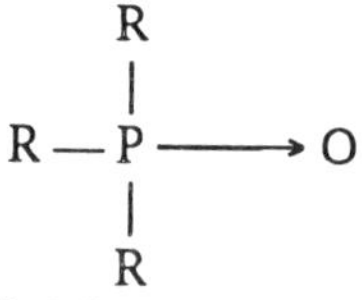

Alkyl dialkyl phosphonate

The organophosphorus compound extract most of the metal from aqueous phase by solvating them.

The most widely used organophosphorus compound is *TBP* (tri-n-butyphosphate). In the extraction of metal nitrates with TBP the most highly charged ions are most readily extracted. If ions are of same charge, then the one with smaller ion i.e. the greater charge density, is readily extracted, i.e., the metals are extracted in the order :

$$Na^+ < Ca^{2+} < Ce^{3+} < Ge^{+5}$$

$$Li^+ > Na^+ > K^+ > Cs^+ \text{ and } Ca^{2+} > Sr^{2+} > Ba^{2+}$$

Because of its importance in nuclear technology, the extraction of $UO_2 (NO_3)_2$ into organophosphorous solvents has been thoroughly investigated.

Rate of attainment of equilibrium is important in extraction system. When equilibrium is not readily established then rate of attainment of equilibrium must be added to all the factors that have been discussed in extraction system. For example certain extraction systems require hours together for complete extraction, i.e., nickel oxinate requires several hours for extraction into chloroform.

Applications

Solvent extraction plays vital role in chemical industries and nuclear technology and also widely used in chemical synthesis for the purification of the product. There is no field of chemistry where solvent extraction is not utilized. Below are given the details of some of the solvent extraction system

Experiment No. 8.1. To study the effect of pH on the extraction of copper-oxinate chelate in chloroform.

Cu^{2+} reacts with oxine to form neutral inner complex according to the reaction.

Cu^{2+} + 2 [oxine: N, OH] ⟶ [N → Cu/2, O — Cu/2] + $2H^+$

The neutral complex is extractable into chloroform. Since during chelation H^+ are produced, extraction will be pH dependent. The equation 932) gives the pH dependence of extraction.

Procedure. Pipette out 1 ml of copper sulphate solution (0.500 mg Cu/ml) in a 250 ml separatory funnel. Add 49.0 ml of distilled water and about 25 drops of 1 M HCl to keep pH ~ 1.5. Pipette 1 ml of 0.1 M oxine in chloroform and 9.0 ml of $CHCl_3$ into the funnel (presaturation of $CHCl_3$ with H_2O are that of water with $CHCl_3$ is not desired as the solubility of each in each other is negligible). Stopper the funnel and shake vigorously, venting frequently for two minutes. Allow the funnel to stand for a minute and again shake for two minutes. Rest again for a minute and shake it again. Allow to stand the funnel, until two phases are clearly separated. Remove the lower layer and re-extract the aqueous phase again with 1 ml of 0.1 M oxine and 9 ml of $CHCl_3$. Mix the two organic phases together. Measure the absorbance of the copper oxinate complex at 450 nm on a spectrophotometer or with the help of a colorimeter. Alternatively, distill of the chloroform layer and finally evaporate nearly to dryness. Then add concentrated HNO_3 and heat strongly to break the complex. Cu^{2+} in aqueous phase can be analysed complexometrically using PAN as the indicator in acetate buffer at pH around 5.0.

Repeat the above procedure nine times, charging only the amount of HCl initially added. Add about 22, 20, 15, 12, 10, 8, 5, 2 and 0 drops of 1 M HCl respectively in these additional extractions.

Plot the % *E* (ordinate) vs. pH (abscissa) (or vs. number of drops of HCl). From the resultant graph calculate $pH_{1/2}$ for Cu-oxinate extraction system. From the plot, determine the pH where % E is 40, 45, 50, 55 and 60 respectively. Calculate *D* for each % *E* and then log *D*. Substitute the values for *D* and pH in equation (31 *a*) and determine *n* and *K'*.

Experiment No. 8.2. Separation and extraction of Cu (II), Pb(II), Mg(II) oxinates from their mixtures and spectrophotometric determination of each ion.

The yellow oxinates of Cu(II), Pb(II) and Mg(II) are extractable into chloroform. The $pH_{1/2}$ values for copper oxinate, lead oxinate and magnesium oxinate are 1.37, 5.04 and 8.57 respectively when extracted in 0.10 M oxine in chloroform. Sine $pH_{1/2}$ values are different for these chelates, a proper adjustment of pH will allow their separation from their mixture. Copper-oxinate is completely extracted at pH 3.0, lead oxinate at pH 7.0 and magnesium oxinate is extracted at pH 9.5. Fig. 8.7 shows their extraction curves. Each of the metal chelates is then determined colorimetrically at appropriate wave length. The wavelengths for maximum absorption of the complexes are Cu = 410 nm, Pb = 400 nm, Mg = 385 nm.

Procedure. Pipette 1.0 ml of aliquot of an unknown solution containing not more than 0.2 mg of each metal ion (at higher concentration yellow precipitates in aqueous phases are formed and that complicates the extraction) in a 250 ml separatory funnel.

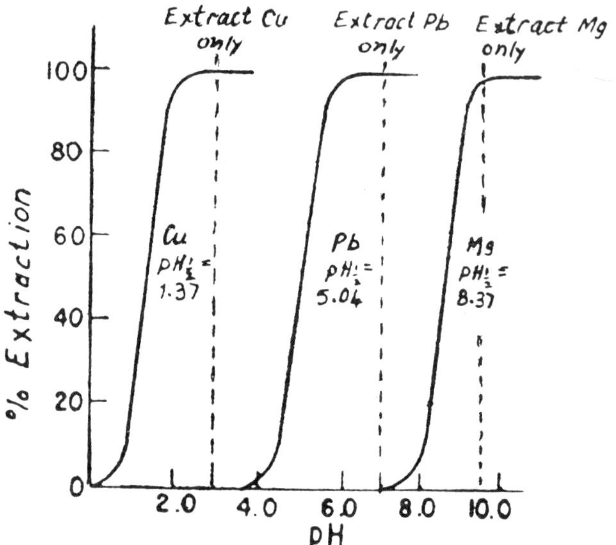

Fig. 8.7. Extractive separation of copper, lead, and magnesium with 0.1 M oxine in chloroform.

Add 10 ml of distilled water and acetate buffer of pH 3.0 (10 ml). Then add 10 ml of 0.1 M oxine in $CHCl_3$. Shake the funnel for two minutes, frequently venting. Allow the funnel to stand for two minutes to separate the two layers. Remove the lower layer carefully in 50 ml measuring flask (allowing precipitate if any in the funnel). Repeat the extraction with 5 ml of 0.1 M oxine solution. Combine both the extracts and make the total volume to 50 m with $CHCl_3$ and measure the absorbance against the oxine blank at 410 nm (within few hours as copper oxinate is not stable over night) or remove $CHCl_3$ by distillation and decompose the complex with HNO_3 (as described in experiment No. 1) and estimate complexometrically.

Transfer most of the aqueous phase to 100 ml beaker. Adjust the pH to 6.0 with the pH meter, using concentrated ammonia and then raise carefully to 7.0. Quantitatively transfer the whole contents (even if precipitates are there) to 250 ml separatory funnel using minimum amount of distilled water. Extract Pb-oxinate in 0.1 M oxine/CHC_3 [according to the procedure followed for Cu (II)] Check the pH. Readjust to 7.0, if necessary. Less than four extractions with 10 ml, 5 ml, 5 ml, will be sufficient for complete extraction. After raising the volume to 5 ml with $CHCl_3$ in a measuring flask, measure absorbance at 400 nm or decompose the complex as described for Cu (II).

If there is no other interfering metal present, the aqueous phase left after extraction of Pb-oxinate can be directly utilized for estimating Mg (II) complexometrically. For studying the complex spectrophotometrically, transfer the aqueous phase to a beaker and raise the pH to 9.5 using strong ammonia solution or use ammonia/ammonia chloride buffer. Shake the solution for less than one minute (since Mgloxinate is very unstable). Allow the phases to separate. Remove the lower layer is 50 ml measuring flask. Reextract with 10 ml portion of 0.1 M oxine in chloroform. Make the extract to mark and measure the absorbance at 400 nm or decompose chemically and estimate as described for Cu(II). From the calculation of each metal in different extracts, amount of each metal can be estimated in the mixture.

Experiment No. 8.3. Extraction of iron as ion pair ($H^+ FeCl_4^-$) in isopropyl ether.

Iron in concentracted HCl medium forms a complex $H^+ FeCl_4^-$, which acts as an ion of the ion pair which is extractable in isopropyl ether (less volatile than ether).

Procedure. To an aliquot of iron (< 0·5 g) in low nitrate content add sufficient HCl to have final concentration of 7·5 to 8·0 M in HCl. Transfer the solution quantitatively to separatory funnel washing with 7·5 M HCl. Add 25 ml of isopropyl ether and shake the funnel for two minutes, venting the funnel frequetly. Allow the two phases to separate. Carefully drain out the aqueous layer in second separatory iunnel after washing the isopropyl layer properly with 7.5 MHCl. Add 25 ml of isopropyl ether and extract the remaining $H^+ FeCl_4^-$ ion pair. Repeat the extraction till isopropyl layer is a colourless. Collect all the organic layers and shake with 25 ml of distilled water. All the iron will be transferred to aqueous layer.

Estimation of iron. To suitable aliquot of the iron solution, add excess EDTA and raise the pH to 5.0. Titrate excess EDTA back with Cu^{2+} using PAN as indicator or estimate Fe Colorimetrically using 1 : 10, phenanthroloine as chromogenic reagent. For this transfer the aquous layer of iron to 1 litre volumetric flask. Add 3 ml of conc. H_2SO_4 and make up to mark with distilled water.

Pipette a suitable aliquot of this solution to have 0.1 to 0.5 mg of Fe into 100 ml volumetri flask. Add 25 ml of pH 4.0, 0.5 M acetic acid—sodium acetate buffer. Add 5% aqucous $NH_2OH \cdot HCl$ solution and 5 ml of 0.25% 1 : 10—phenanthroline solution in water. Dilute to the mark and mix well. Measure the absorbance at 515 nm.

Fe (III) can be reduced to Fe(II) with $SnCl_2$ (and $SnCl_2$ can be removed by adding $HgCl_2$). Fe (II) can be titrated with K_2Cr_2O. using diphenyl amine as the indicator.

Experiment 8.4. Determination of Pb(II) by extracting lead dithizonate.

The primary metal dithizonates are formed according to the reaction.

$$M^{n+} + nH_2Dz \rightleftharpoons M(HDz)_n + nH^+$$

where H_2Dz ≡ dithizone ≡ $HS{-}C\begin{matrix} /\!\!/ N{-}NHC_6H_5 \\ \diagdown N = NC_6H_5 \end{matrix}$

Some metals such as Cu(II). Ag(I), Au(III), Hg(II), Bi(III) Pd(II) from secondary complexes of the type M_2Dz_n at higher pH values. Only primary metal dithizonates are of analytical importance.

Procedure. Dissolve 0.0067 g of pure $PbCl_2$ in 1 litre of water graduated flask. Take 10 ml of the aliquot (~ 50 μg Pb(II) in a 250 ml separatory funnel. Add to this about 70 ml of or adjust the pH to 9.5 with the help of pH-meter. Add 7.5 ml of 0.015% dithizone in chloroform [1 ml of this dithizone solution ≡ 20 μg of Pb(II)], then add 17.5 ml of $CHCl_3$ solution. Shake for one minute and allow the two phases to be separated out. Measure the absorbance of the chloroform layer, against the dithizone blank in $CHCl_3$ layer, at 510 nm. Repition extraction can be carried out to check the complete extraction of Pb(II) is chloroform (zero absorbance).

Note. For using dithizone as complexing agent, the various reagents used should be of high purity because dithizone is very sensitive reagent.

Experiment 8.5. Determination of nickel as dimethylgyoxime complex in $CHCl_3$.

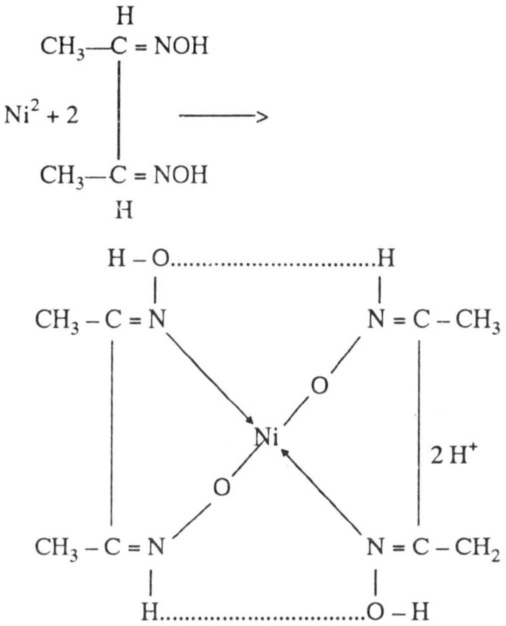

H
CH_3—C = NOH
Ni^2 + 2 | ———>
CH_3—C = NOH
H

H – O..............................H
| |
CH_3 – C = N N = C – CH_3
O
Ni 2 H^+
O
CH_3 – C = N N = C – CH_2
| |
H..............................O – H

Ni $(DMG)_2$

Procedure. Weight out 0.135 g of $NiSo_4$ $(NH_4)_2SO_4$ $6H_2O$ and dissolve in 1 litre of water in a graduated flask. Transfer 100 ml of this solution (Ni ~ 200μ g) to a beaker containing 90 ml of H_2O and 5 g of citric acid (*AR*) and then add dilute ammonia to raise the pH to 7.5. Cool and transfer to separatory funnel. Add 20 ml of 0.1% dimethylgloyxime in 1 : 1 ammonia solution), after standing for two minutes add 12 ml of chloroform solution. Shake for 1 minute, allow the phases to separate. Drain out the chloroform layer. Measure absorbance at 366 nm against the *DMG* blank in Chloroform. A second extraction with 12 ml chloroform will give very little nickel or no nickel.

DMG reagent is prepared by dissolving 0.5 g of *DMG* in 250 ml of ammonia and then diluting it to 500 ml with water.

Application of solvent extractions are unlimited and each and every application cannot be discussed. For detailed studies the references given at the end can be consulted.

REFERENCES

For $H^+ FeCl_4^-$ extraction see R.W. Dodson, G.J. Forney and E.H. Swift, J Amer. Chem., Soc., 58, 2407 (1936) : M. Barrachima and M.A. Villar, J. Inorg. Nucl. Chem, 28, 2407 (1966).

For Cu-oxine extraction : J Stary, Anal, Chim, Acts, 28, 132 (1963).

For continuous counter current : J.J. Kolfenbach, E.R. Kooi et. Al., Ind. Eng. Chem. Anal. Ed., 15,473 (1944).

Solvent extraction reviews : ''Treatise on Analytical Chemistry, Part I, Vol. 3, 1961 by Irving and Will iams and Progress in Inorganic Chemistry, Vol. 2, 1960 by Diamond and Tuck.

Alders. Liquid-liquid extraction. Elsevier Publication, 1955.

Morrison and Freiser. Solvent extraction in analytical chemistry Wiley, 1957.

D e, Separation of heavy metals Pergamon Press, 1961.

Stary, Solvent extraction of metal chelates, Pergamon Press, 1974.

Vogel, A text book of quantitative inorganic analysis. ELBS, 1978.

R.A. Day and A.L., Underwood, Prentice-Hall, 1977.

PROBLEMS

1. Distinguish between partition coefficient and distribution ratio.
2. You are required to preform the extraction in two ways–
   (*a*) Do a single extraction using 100 ml portion of ether or
   (*b*) divide the ether into two 30 ml portion and perform two successive extraction ? Which method is better and why ? Explain it.
3. Derive a relation for *D* which correlates the ability constant of the complex, dissociation constants of the chelon and partition coefficients of metal chelate and chelon between organic layer and water.
4. The value of D for a solute (organic : aqueous) is 10.0.
   (*a*) If 100 ml of water containing 1.0 g of *A* is shaken with 100 ml of ether then what percentage of *A* is extracted into ether ?
   (*b*) If aqueous phase is extracted twice with 50 ml of ether, what percentage of *A* is extracted ?
5. 100 ml of a solution which is 0.100 M in weak acid *HA* is extracted with 25 ml of ether. After the extraction a 25.0 ml aliquot of the phase required 20.0 ml of 0.0500 M NaOH for complete titration. Calculate the value of *D* for *HA* (organic to aqueous).
6. When aqueous solution of ferric chloride in HCl is shaken with twice its volume of ether containing HCl, 99% of the iron is

extracted, calculate the value of *D* (organic to aqueous) of the compound.

7. Two extractions with 20 ml portion of an organic solvent removed 89% of a solute from 100 ml of an aqueous solution. Calculate the distribution ration (organic ; aqueous) or the solute.
8. The partition coefficient of *X* between $CHCl_3 : H_2O$ is 9.5. Calculate the concentration of '*X*'' remaining in the aqueous phase after 50 ml of 0.15 M, of *X* is treated with the following quantities of chloroform :
   (*a*) one 40 ml portion,
   (*b*) two 20 ml portions,
   (*c*) four 10 ml portions and
   (*d*) eight 5.0 ml portions.
9. What total volume of $CHCl_3$ would be required to decrease the concentration of *X* to 1.0×10^{-4} M, if 25 ml of 0.05 M of X were extracted with
   (*a*) 25.0 ml portion of solvent,
   (*b*) 10.0 ml portions of solvent and
   (*c*) 2.0 ml portions of solvent ?
10. What would be the minimum partition coefficient that would permit removal of 99% of a solute from 50 ml of water with
    (*a*) two 25 ml extractions with benzene and
    (*b*) five 10 ml extraction with benzene.
11. 0.00396 g of Br_2 is dissolved in 50 ml of water, this solution is shaken with 10 ml of CCl_4. Calculate the number of milligrams of bromine present in each of the layers at equilibrium, K_D for bromine (CCl_4 H_2O) is 27.0 at 25°C.
12. The partition coefficient for phenol between i-octanol and water is 28.8. Calculate the percentage of phenol remaining unextracted when 0.146 g of phenol, dissolved in 125 ml of water is extracted once with a 15 ml portion of 1-octanol.
13. Calculate the percentage of oxine extracted into an equal volume of aqueous solution from 0.1 Mioxine in chloroform at a pH of
    (*a*) 200 (b) 7.,00 (c) 12.0
14. A weak acid *HA*., has $K_{D_{HA}}$ (benzene HCO = 4.75 and K_a (aq) = 3.91 = 10^{-4}. The acid does not associate in benzene phase. Calculate the pH at which 50% of this acid will be extracted from water into benzene.

15. HCOOH has $D = 0.416$ (methylisobutyl ketone/H_2O) at pH 2.13. Assuming no association of HCOOH in organic layer. Calculate. $K_{D_{HCOOH}}$ (benzene/H_2O).

16. Copper extracts into 0.1 M oxine in $CHCl_3$ with $pH_{1/2} = 1.38$. If the $K_{D_{oxine}}$ ($CHCL_3/H_2O$) = 433 and that for copper-oxine in the same system is 3.02×10^3. Calculate the overall stability constant for $Cu_{(oxine)_2}$.

17. The following metal ions are extracted into 0.1 M benzoylacetone in benzene with indicated extraction constants. Calculate the $pH_{1/2}$ value for each of these extractions, considering the formula for the chelate in each case is $M(BzA)_2$

Metal ion	log K
Cd(II)	—14.11
Co(II)	—11.11
Cu(II)	— 4.17
Ni(II)	—12.12
Pb(II)	— 9.61

From the calculated $pH_{1/2}$ values estimate which of the following pairs of metal ions can be separated by extraction with 0.1 M benzoylacetone is benzene and which cannot be separated by this : Lead(II) : copper(II) : nickel(II) : cobalt(II) : lead(II) : cadmium(II). Copper(II) : nickel(II) and cadmium(II) : nickel(II).

18. Zn(II) is extracted with 1×10^{-4} M dithizone in CCl_4 with a $pH_{1/2}$ of 2.80. Calculate the pH at which 99.9% of zinc in an aqueous solution will be extracted with dithizone.

19. Draw the diagram and explain the working of continuous extractors
    (*a*) extracting liquid is heavier than the solution being extracted.
    (*b*) extracting liquid is lighter than the solution being extracted,
    (*c*) solid is extracted by a solvent lighter or heavier.

20. Give the characteristics of good extracting solvent.

21. Discuss the effect of various factors which affect the extraction of metal chelates into organic phase from aqueous phase.

22. Derive the equation

$$W_{ag} = \left(\frac{V_{ag}}{DV_{org} + V_{ag}}\right) n$$

for batch extraction in times. The various terms used have usual significance.

23. Describe the mode of operation of separation process in discontinuous counter current distribution *i.e.*, Craig's counter current distribution operation.
24. How are the extraction systems classified ? Name each of them and discuss each of these in brief.
25. Show that distribution coefficient for a metal chelate of metal M with ligand HX is given by the expression.

$$D = \frac{K_{D,\,HX} \cdot K_{st}\, K_a{}^n [HX]^n{}_{org}}{K^n{}_{D,\,HX}[H^+]^n}$$

where the various constants have usual significance.

CHAPTER 9

General Remarks on Gravimetric Methods of Analysis

Perhaps the most important of the classical methods of analysis are the gravimetric methods which are based upon obtaining the weight of a precipitate. In general, a gravimetric method of analysis consists of the following steps :

(*a*) the sample weight is determined and it is dissolved in a proper solvent.

(*b*) a sequence of chemical reactions is carried out to obtain a suitable product

(*c*) the weight of the product of the reactions is determined.

(*d*) this weight in conjunction with the knowledge of the product is used to calculate the quantitative composition of the sample.

The key chemical reaction involved in gravimetric methods is almost always a precipitation, because it can be a highly selective means for separating the component of interest from the matrix, and the product weighed is almost always a precipitate, because it can be weighed readily.

Applicability of Gravimetry and its Advantages. Eminently practical considerations of time and economics often dictate the choice of a particular type of analytical method. Rarely can a flowing process stream in a plant be stopped while a batch sample is removed and analyzed gravimetrically. Consideration of the time factor must also include consideration of the present-day high cost of labour. Otherwise the method is selected by such considerations as the needed precision and accuracy and the required selectivity.

Like titrimetery, gravimetric analysis also requires relatively inexpensive, simple, maintenance free equipment. Suitable selective procedures which are applicable to a wide variety of sample types with good precision and accuracy are available. Gravimetry has the virtue of giving reliable results in terms of isoated products whose purity and identity can be confirmed, if necessary.

It may be true that fewer people now have the board background of descriptive solution chemistry and the practiced skills of the experienced wet analyst, but the time required to acquire such skills is not appreciably greater than that needed for the intelligent use of many specialized physicochemical techniques. The time consuming nature of operations as weighing, evaporation, filtration, and ignition can be greatly reduced by proper planning.

Generally, one expects that gravimetric method will be applicable for constituents falling within the concentration range of 1 to 99% and with a relative accuracy ranging from at least ± 1% to about ± 0.1%. With minor (< 1% constituents, other methods will often be preferable or more rapid.

Within the limitations just discussed, gravimetry is attractive, then, in view of its minimal apparative requirements and its ready applicability for most elements in most types of matrices, with good precision and accuracy over the concentration range of commonest significance. Inspite of instrumentation and automation, gravimetric methods have not lost their importance. It could be rather difficult to find a instrumental method which does not need the methods of classical chemical analysis to provide to composition of the standard used for calibrating the instrument. Gravimetry will probably long remain an important approach for the analysis of isolated samples of diverse nature.

General Requirements of a Gravimetric Method. For gravimetric analysis, the precipitation reactions should have the following criteria :

1. It should precipitate only the compound of interest.
2. The precipitation should be quantitative.
3. The particle size of the precipitates should be large enough so that it could be washed, filtered or dried with little difficulty.
4. The precipitate should be a compound of known composition or it should be easily converted into such a compound.

General Requirements of Precipitate. The Precipitate must either itself be a stoichiometric compound that is easy to weight, it should be nonvolatile, nonhygroscopic, nonefflorescent, and inert to reaction with air, or it should be easily convertible into such a compound by drying or ignition.

The precipitates which are easily separable from the medium in which they have been formed are required in gravimetric methods. These conditions are satisfactorily fulfilled by quite a large number of precipitates, even when the selectivity of action implied by the word "pure" is recognized.

Quantitative analysis, however, adds further very stringent requirement that such a purity he maintained simultaneously with the complete and quantitative removal of a given ion. Its idealized aim is no less than that of obtaining a 100% recovery of a 100% pure product. This becomes an entirely different matter, and results in the existence of a rather finite number of really good gravimetric methods. This dilemma is easily illustrated by considering some of those factors and conditions which would usually operate to improve the purity of a precipitate :

1. High dilution of the sample solution.
2. Use of dilute solutions of reagents.
3. Presence of minimal excesses of any common ions.
4. Precipitation at elevated temperatures.
5. Absence of organic solvents, which may lower solubility of contaminants.
6. Minimal contact of precipitate formed with mother liquor, if post precipitation is possible.
7. Extensive washing of the precipitate.

It is obvious that each of these conditions represents the opposite o the requirement for minimizing the solubility losses. The desire for easily filterable particles is generally more compatible with the conditions for purity.

This explains the compromises which are always necessary, and the need for prescribing the best of these based on actual trials. It justifies the frequent practice of forming and digesting the bulk of a precipitate at an elevated temperature, but of cooling it to a lower temperature prior to filtration.

Above all, it emphasizes the desirability of adhering to the prescribed procedural steps if the prescription is expected to achieve the intended results.

Separation of Precipitates. The solid-liquid phase separation may be achieved by decantation, filtration centrifugation, or a combination of these procedures.

The details of a separation by filtration will depend very greatly on the nature of the precipitate, primarily on whether it is crystalline in character or more or loss colloidal. The nature and composition of the apparatus used must also be suitable for the intended low-temperature drying or high temperature ignition, and chemically resistant to the precipitating medium Crystalline materials can be filtered in various types of filter crucibles, but colloidal precipitates are usually collected on filter papers for ignition and weighing in ordinary crucibles. The dimensions of filtering devices should be dictated more by the mass of the precipitate than by the volume of liquid involved. Siphon and suction devices facilitate the handing of larger volumes of liquids.

The pore diameter in a filter medium should be below the minimum dimensions of the particles to be filtered. If the pore size is too large, turbidity will appear in the filtrate (not to be confused with the peptization of colloids), and in intermediate cases the pores become obstructed,, slowing the filtration. If the pore size is too small, the filter medium may clog and filtration may be unduly slow.

Contamination of precipitates. Contaminations may be classified into two groups : those due to co-precipitation and those caused by post-precipitation. In co-precipitation, two types of contamination can be distinguished : adsorption and occlusion. In the case of adsorption, the precipitate particles carry the contaminant on their surface and can be removed by washing. In occlusion, the impurity is inside the primary precipitate particles, whether by mechanical trapping during crystal growth or by formation of solid solutions.

Washing of the precipitate. Precipitates are washed to remove the absorbed impurities, or to replace them with species which can be removed by heating.

Errors which may result from improper washing include incomplete washing, solubility losses, and peptization of colloids.

Tests to determine the completeness of washing are necessary. A sample of the filtrates collected and a simple qualitative test for one of the ions in the original mother liquor is applied.

In order to decrease the solubility losses, it is desirable to use the minimum practicable volume of wash solution. For a given total amount of wash solution, it is most effective to wash many times, using a small volume each time, and draining the precipitate as dry as possible after each washing. A dilute solution of a common ion, or a solvent, such as alcohol or acetone, in which the precipitate is very insoluble may also be useful in minimizing the solubility losses.

Pure water is rarely used as a wash solution, both because solubility losses are maximized in pure water, and because water tends to remove all ions, which in case of colloidal precipitates causes peptization. Therefore, if the precipitate shows any tendency peptize, a wash liquid which contains electrolyte is used, and with sufficient washing the ions of the wash solution replace the originally adsorbed species. Dilute solutions of ammonium salts are most common wash liquids.

The choice of any component present in the wash liquids will be guided, of course by the following objectives :

1. They may repress the solubility of the precipitate.
2. They may favour the removal of expected contaminants.
3. They may function to discourage the peptization of the colloidal precipitates.
4. They should be relatively volatlie or decomposable acids, bases or salts.
5. Their presence in the filtrate should not unduly complicate any subsequent separations.

The actual choice of such a material then, is often a rather specific one for a given circumstance.

Whenever possible, the last few washes will consist of pure water or solvent. Some precipitates are directly weighable after suctioning dry in a filter crucible. The final washes in these cases may consist of ethanol, or of ethanol followed by ether.

Drying and Ignition. Once obtained, collected, and washed with a suitable liquid, the precipitate is freed from water by drying or ignition. During this heat-treatment the precipitate may undergo physical or chemical changes which influence its weight and composition.

Winkler has drawn attention to the fact that crystalline precipitates can be dried to constant weight even at room temperature if they are not hygroscopic, do not tend to oxidize, do not react with carbon dioxide present in air and do not decompose in air, though naturally this may take a long time. Sometimes water is removed with alcohol etc.

The precipitates which possess the necessary properties may be directly weighed after drying or it might be weighed after ignition to give a compound of a known composition. Normal precautions should be followed during the weighing of the precipitate.

Precipitation from Homogeneous Solution

One of the major objectives in quantitative precipitation in gravimetric analysis is the separation of a *pure solid phase*. This may be achieved by precipitation from solution in which the supersaturation is as low as possible. The importance of small degree of supersaturation was for the first time, indicated by van Weiman—when the supersaturation is small fewer nucki are initially formed ; these nuclei then grow to large aggregates as additional solid phase deposits on them. The product thus obtained is more easily filtered and is less contaminated by coprecipitation.

It is to maintain small supersaturation that analytical chemists follow the practice of adding a dilute solution of a precipitating regent slowly and with stirring. Willard and Teng, for example, were able to precipitate 100 mg of Al in exceptionally dense form by adding 0.002 M ammonia solution over a period of several days. The same authors also found that the result could be obtained by *in situ* generation of OH^- by hydrolysing urea in an hour or less in the presence of Al^{3+}. The obvious practical limit to these expedients are the increased solubility losses in the more dilute solutions and tedium of slow addition process. Furthermore, these techniques do not entirely eliminate high supersaturation and rapid precipitation. At the point in solution a new drop of the reagent first makes contact, there is atleast momentarily a high degree of concentration of the reagent. Until stirring disperses the drop there exists a high local supersaturation and consequent rapid formation of the solid. Even with the most efficient working this local effect cannot be eliminated.

"Precipitation from Homogeneous Solution" is an ingenious method for introducing a precipitating reagent in such a such a way as to avoid the local excesses that accompany mechanical addition of the reagent. The precipitation is caused by slowly generating the reagent throughout the entire solution by means of a suitable chemical reaction. During the process the solvent remains homogeneous with respect to the concentration of the ions being precipitated and the reagent ; if the generating reaction is slow high supersaturation is avoided and gradual formation of the solid takes place throughout the entire solution.

Since the time this valuable technique was recognized many improvements have been made in its application. The method is no longer continued to getting easily filterable and least contaminated precipitate but has been found useful in the separation of such ions which are otherwise difficult not impossible to separate.

The various methods for precipitation from homogeneous solution could be divided into the following classes :

(1) The direct generation *in situ* of the precipitation reagent by a controlled reaction.

(2) A gradual change of pH to decrease the solubility of the desired compound slowly.

(3) The gradual removal of the solvent.

(4) The break-down of a suitable complex to produce an insoluble substance ; this process can take two forms : (*a*) Complete chemical breakdown of the complex by a process such as oxidation (*b*) Conversion of the complex into a more stable one accompanied by the liberation of the ion being investigated so that it is now able to take part in precipitation reaction.

The above methods are briefly described below :

(1) **The direct generation of the precipitating agent.** A summary of the important ions/reagents, their sources and the reactions showing their formations and important application is presented below :

Anions	Source	*Reactions*
OH^-	Urea	$(NH_2)_2CO + H_2O = 2NH_3 + CO_2$
OH^-	Sulphamic acid	$HNH_2SO_3 + H_2O = NH_4^+ + H^+ + SO_4^{2-}$
S^{2-}	Thioacetamide	$CH_3CSNH_2 + H_2O = CH_3CONH_2 + H_2S$
CO_3^{2-}	Trichloracetate	$2\,CCl_3COO^- + H_2O = 2CHI_3 + CO_2 + CO_3^{2}$
$C_2O_4^{2-}$	Dimethyloxalate	$(CH_3)_2C_2O_4 + 2H_2O = 2CH_3OH + H_2C_2O_4$
PO_3^{3-}	Trimethylphosphate	$(CH_3)_3PO_4 + 3H_2O = 3CH_3OH + H_3PO_4$
$C_1O_4^{2-}$	Urea and $HCrO_4^-$	$2HCrO_4^- + (NH_2)_2CO + H_2O = NH_4^+ + CO_2 + 2CrO_4^{2-}$
IO_4^-	Acetamide and H_5IO_6	$H_3IO_4 + 5CH_3CONH_2 + 3H_2O = 5CH_3COONH_4 + H^+ + IO_4^-$

Urea has been used in the precipitation of Al, Ga, Fe, Th, Sn, and Zr. Other anions sometimes affect the precipitation. For example. Al can be precipitated in presence of SO_4^{2-}, succinate or formate but not in the presence of Cl^-, NO_3^- or acetate. The presence of a suitable anion has been demonstrated to be due to the formation of a basic salt in forming a dense precipitate.

PO_4 generated from trimethylphosphate has been used in determination of Zr. Triethylphosphate and metaphosphoric acid are also used to generate PO_4^{3-} ion slowly.

$C_2O_4^{2-}$ ion on being generated slowly is useful in precipitating Ca, Th, and rare earths. In addition to diethyloxalate is equally effective in these precipitations.

SO_4^{2-} obtained from the hydrolysis of sulphamic acid has been used in the separation and determination of barium. Dimethyl sulphate found to be equally effective in the determination of Ba, Sr, Ca and Pb.

S^{2-} from acetamide has limited application because of the contamination of the precipitate with sulphur. It has been applied in the determination of SB, Bi, Mo, Cu, As, Cd, Pb, Sn, and Hg.

CO_3^{2-} generated from trichloroacetic acid has been used in the determination of Nd.

CrO_4^{2-} generated from the slow oxidation of Cr(III) with bromate has been applied in the determination of Pb.

IO_4^- has been used for the determination of thorium.

Reagents. Like anions, certain reagents are generated in the reaction medium causing precipitation from homogeneous solution of the ion. Some examples of these are given here.

The slow generation of 8-hydroxyquinoline by hydrolysis of 8-acetoxyquinoline is used for Al, Mg, Th, U. etc.

Salicylaldoxime is produced by the reaction between salicylaldehyde and hydroxylamine hydrochloride. The method is useful for Cu.

Cupferon is generated from the reaction of 2-naphthol with nitrous acid. The reagent is used for estimating Co.

Dimethylglyoxime is obtained when biacetyl reacts with hydroxylamine. Ni and Pd are estimated in this way.

Dihydroqueretin when boiled is converted into quercetin and is used for Nb and Ta.

(2) **A gradual change of pH.** Urea hydrolysis is the most common method for increasing the pH. The method has been found useful in the precipitation of the hydroxides of Fe and Al, hydroxyquinolates of Mg and Nb, and basic formate of Bi.

(3) **Removal of solvent.** An ion and organic reagent are mixed in a suitable water miscible solvent and the solvent is then evaporated slowly. In the precipitation of Al-hydroayquinoline acetone is allowed to evaporate from the 50 : 50 acetone—water mixture effecting the precipitation from hom geneous solution.

(4) **Break-down of complexes.** Several methods are there for the break-down of complexes by which the precipitation from homogeneous solution of various metal ions can be effected. These are briefly outlined here :

(*a*) **Heat.** Tungsten in nitric acid and hydrogen peroxide forms peroxo acid of tungesten which can be decomposed by heat precipitating tungstic acid. Other examples of similar type are known.

(*b*) **Oxidation.** EDTA complexes are generally broken resulting into the slow release of metal ions. For example the precipitation of hydrous iron oxide released from the EDTA complex by slow oxidation with H_2O_2 gives dense and easily filtrable precipitate.

(*c*) **Replacement of one ion by another.** Several estimations of this type have been done by the present author. Replacement of thallium from its EDTA complex by magnesium and its precipitation as chromate has been found to be an excellent method. Replacement of copper from its EDTA complex by causing its reaction with ascorbic acid is found to give excellent precipitate in its determination as copper(I) thiocyanate. Determinations of Zn, Bi, and a few other metals have been carried out successfully by ion release. The ion release method not only gives very good filtrable precipitate but also avoids interference by many ions.

The technique of precipitation from homogeneous solution. PFHS as it is called, is indeed an excellent technique in gravimetric estimations. However, corresponding to movements in mechanics of filtration and washing, the reduction in contamination has not always been significant.

CHAPTER 10

Redox Titrations

Definitions

Oxidation refers to any chemical reaction in which oxidation number increases ; *reduction* refers to a chemical change in which oxidation number decreases. Thus, oxidation occurs with loss of electrons and reduction is accompanied by a gain of electrons. For example in the oxidation of Fe^{2+} with acidified $KMnO_4$ $MnO_4^- + 8H^+ + 5Fe^{2+} \rightarrow Mn^{2+} + 5Fe^{3+} + 4H_2O$

MnO_4^- (permanganate) undergoes a decrease in oxidation number, thus it is reduced. The oxidation number of Fe^{2+} increases from 2 to 3 ; thus it is oxidised.

Oxidising Agent is one in which one atom (such as Mn in $KMnO_4$) undergoes a decrease in oxidation number. Thus the oxidation number of Mn in $KMnO_4$ decreases from +7 (in MnO_4^-) to +2 in Mn^{2+}

Reducing agent is a substance whose one atom undergoes an increase in oxidation number. Thus in Mohr's salt the oxidation number of Fe^{2+} increases from 2 to 3.

Oxidation and reduction always occur together and compensate each other. The term oxidising or reducing agent refers to the entire substance and not to just one atom. Some reagents act as both oxidising or reducing agents. Under different conditions, e.g. H_2O_2.

Some Common Oxidising Agents

Examples of various oxidizing agents used in redox titrimetry include potassium permanganate, potassium diehromate, salts of quadrivalent cerium, pentavalent vanadium, trivalent iron, potassium iodate, potassium bromate, etc.

(1) Ceric sulphate is very good oxidising agent used with O-phenanthroline as an indicator.

$$Ce^{4+} \rightarrow Ce^{3+} + e^-.$$

Ce (IV) during reaction exists as an anionic complex in acidic medium (H_2SO_4 , HNO_3 and $HClO_4$). The formal potential of Ce (IV)-Ce (III) couple is 1.70 in $HClO_4$; 1.60 V in HNO_3 and 1.42 V in H_2SO_4 solution. $Na_2C_2O_4$ is used as the primary standard for cerium (IV) in sulphuric acid. Ammonium hexanitro cerate, $(NH_4)_2Ce(NO_3)_6$, is used for the preparation of standard cerium (IV) solution. Ce (IV) is used mainly in titrations of Fe (II), with ferroin as indicator in 0.5-8 M H_2SO_4 or $HClO_4$ or 0.5-3 M HCl medium. H_2O_2, Oxalates. As (III), ferrocyanide can be determined by direct titration. Some of them require catalyst e.g. the titration of $H_2C_2O_4$ against Ce (IV) in 0.5-2 N HNO_3 employs $AgNO_3$ as catalyst-to complete the reaction. Ce (IV) has as advantage over $KMnO_4$—the use of high concentrations of HCl as the reducing medium because Ce (IV) fdrms a stable complex. The solution does not oxidise in dilute hydrochloric acid (0.5-3 M).

(2) *Potassium permanganate.* It is a powerful oxidising agent, and no indicator is required, it acts as its own indicator. Its tendency to oxidise chloride ion and limited stability of its solutions are the only limitations. It is employed in acid (0.1 N) media.

$$MnO_4^- + 8H^+ + 5e^- \rightleftharpoons Mn^{2+} + 4H_2O ; \quad E° = 1.51 \text{ V}$$

The oxidation reaction is rapid. For $H_2C_2O_4$ it is slow at room temperature ; hence heating is required. Reaction with As (III) needs a catalyst. The permanganate end point is not permanent and slowly fades due to the reaction

$$\underset{\text{coloured}}{2MnO_4^-} + 3Mn^{2+} + 2H_2O \rightleftharpoons \underset{\text{colourless}}{5MnO_2} + 4H^+$$

Aqueous solution of $KMnO_4$ is not stable and oxidises water in as follows :

$$4MnO_4^- + 2H_2O = 4MnO_2 + 3O_2 + 4OH^-$$

The decomposition is catalysed by light, heat, acids, bases, Mn (II) ion and MnO_2. In the preparation of a standard solution of $KMnO_4$ the presence of MnO_2 should be avoided. The solution is standardized against $Na_2C_2O_4$

$$2\,MnO_4^- + 5\,H_2C_2O_4 + 6\,H^+ \rightleftharpoons 2\,Mn^{2+} + 10CO_2 + 8\,H_2O.$$

(3) *Potassium dichromate* : The half reaction can be represented as :

$$Cr_2O_7^{2-} + 14\,H + 6\,e^- \rightleftharpoons Cr^{3+} + 7\,H_2O \quad E^\circ = 1.33\text{ V}$$

$K_2Cr_2O_7$ has limited application in comparison with $KMnO_4$ or Ce (IV). This is due to the weaker oxidising characteristics and slowness of some of its reaction ; $K_2Cr_2O_7$ is very stable and inert towards HCl. It can be obtained in high purity and is a primary standard. Diphenylamine sulphonic acid is used as the indicator for $K_2Cr_2O_7$ oxidations $K_2Cr_2O_7$ is mainly used for the analysis of Iron (III) :

$$6\,Fe^{2+} + Cr_2O_7^{2-} + 14\,H^+ \rightleftharpoons 6\,Fe^{3+} + 2\,Cr^{3+} + 7\,H_2O$$

(4) *Potassium bromate* is strong oxidising agent.

$$BrO_3^- + 6\,H^+ + 6\,e^- \rightleftharpoons Br^- + 3\,H^2O \quad E^\circ = 1.44\text{ V}$$

In acidic solutions BrO_3^- reacts with Br^-. $KBrO_3$ is a primary standard and is very stable. Potassium bromate finds extensive application in organic chemistry.

(5) *Potassium iodate*

$$IO_3^- + 5\,I^- + 6\,H^+ \rightarrow 3\,I_2 + 3\,H_2O$$

KIO_3 is extensively used in malytical chemistry and in Andrew's titration

$$IO_3^- + Cl^- + 6\,H^+ + 4\,e^- \rightarrow ICl + 3\,H_2O \quad E^\circ = +\,1.20\text{ V}$$

The Andrew's titration is carried out with 6 M hydrochloric acid in the presence of carbon tetrachloride. The end point is indicated by disappearance of violet colour. Vigorous shaking is necessary for getting an accurate end point.

(6). *Iodine.* It can act both as an oxidising as well as a reducing agent depending upon conditions. The E° of the codine—codide couple is represented by the equation:

$$\frac{1}{2}I_2(s) + e^- \rightarrow I^-(aq) \quad E^\circ = +\,0.535\text{ V}$$

Iodine is however, a weak oxidant and weak reductant. It is sparingly soluble in water and its solutions are generally prepared in KI solution, in which it dissolves forming the tritodide, KI_3. Iodine is not a primary standard and hence it must be standardised against of AS_2O_3 as primary standard. The loss due to volatility and air oxidation of iodide promotes error in analysis. The standardisation can also be carried out against $Na_2S_2O_3 \cdot 5H_2O$. Thiosulphate is standardised against $K_2Cr_2O_7$ as the former is not a primary standard.

$$Cr_2O_7^{2-} + 14H^+ + 6I^- \rightarrow 3I_2 + 2Cr^{3+} + 7H_2O$$

Usually starch is used as an indicator. Iodide in concentration $< 10^{-5}$ M can be easily detected by starch. The iodine starch complex has limited water solubility and interfere hence it should be added at end point of reaction.

Iodine solution in KI can also be titrated with $Na_2S_2O_3$

$$I_3^- + 2S_2O_3^{2-} \rightleftharpoons 3I^- + S_4O_6^{2-}$$

Excess KI reacts with Cu (II) to form CuI and release in equivalent amount of I_2 according to the reaction :

$$2Cu^{2+} + 4I^- \rightarrow 2CuI + I_2 ;\ 2Cu^{2+} + 3I^- \rightarrow 2CuI + I_3^-$$

Here iodide ion serves as reducing agent.

Best results are obtained with 4% KI. Optimum pH is 4.0. In alkaline medium basic copper (II) hinders oxidation. Slow addition of $Na_2S_2O_3$ is useful as adsorbed iodine is released only slowly. The presence of chloride ion should be avoided as iodide will not reduce Cu(II) quantitatively.

Detection of End point

The end point of a redox titration can be located either by using a visual indicator or by using a potentiometric technique. The visual indicators are called in this case, redox indicators.

Substances which have different colours in the oxidized and reduced states are usually employed as redox indicators. Examples include diphenylaminesulphonic acid, diphenylbenzidine, ferroin, methyl orange, methyl red, triphenylmethane dyes, phthalocyanines, etc.

A good redox indicator should satisfy the following requirements :

(*i*) should have a sharp transition in color at the equivalence point.

(*ii*) the colour change at the equivalence point should preferably be reversible.

(*iii*) the colour change of the indicator should occur at a certain value of the potential, called the transition potential characteristic of the indicator which lies in between the redox potentials of the two couples involved (e.g., the transition potential of ferroin is 1.0 V which lies in between 1.44 of Ce^{4+}/Ce^{3+} and 0.77 V of Fe^{3+}/Fe^{2+} and therefore is an excellent indicator for the titration of Fe^{2+} with Ce^{4+}).

(*iv*) the reduction of the oxidized form of the indicator by the reductant (e.g., ferrin by Fe^{2+}) and oxidation of the reduced form of the indicator (ferroin) by the oxidant (Ce^{4+}) should be instantaneous.

and (*v*) the indicator should not consume any significant quantities of either of the reactants so that the indicator correction is negaligible.

Location of the end point by the potentiometric technique. The ability of a substance to lose or gain electrons (to get oxidized or reduced) varies with the chemical nature of the substance. A substance which can gain electrons with ease (e.g., $KMnO_4$) is a powerful oxidizing agent while others such as I_2 and Fe^{3+} are weak oxidizing agents. The physical quantity which gives a quantitative measure of the ability of an oxidizing agent to gain electrons or a reducing agent to lose electrons is the oxidation-reduction potential or simply redox potential. The equivalence point of a redox reaction can therefore be ascertained by measuring the change in this quantity (redox potential) with a change in the concentration of the reactants. This measurement is possible by setting up a test cell consisting of a bright platinum strip or rod dipping into the solution to be titrated and contained in a suitable titration vessel serving as the indicator electrode and a reference electrode (either standard hydrogen electrode or a saturated calomel electrode) connected through a salt bridge (an inverted 'U' tube filled with an agar-agar jelly containing saturated potassium chloride solution of ammonium nitrate solution) to eliminate the liquid junction potential. The indicator electrode is connected to the +ve and the reference electrode to the -Ve terminals of the potentiometer. The tirtation is then accomplished by adding either the oxidizing or the reducing agent as the case may be, in small instalments to the solution to be titrated and noting the change in potential each time. At the equivalence point, a sudden inflection in potential will be observed and the titration is continued there in potential will be observed and the titration is continued thereafter adding a little more of the titrant, and nothing the potential change. A graph is plotted with the E.M.F. data (versus reference electrode) on the ordinate axis and the volumes of the titrant on the abscissa. Bigger the difference in the redox potentials of the two couples, larger is the inflection potential at the equivalence point. The volume of the tirtant corresponding to the midpoint of inflection in the curve is the equivalence point of the titration. The profile of a typical potentiometric titration curve is given in Fig. 10.1

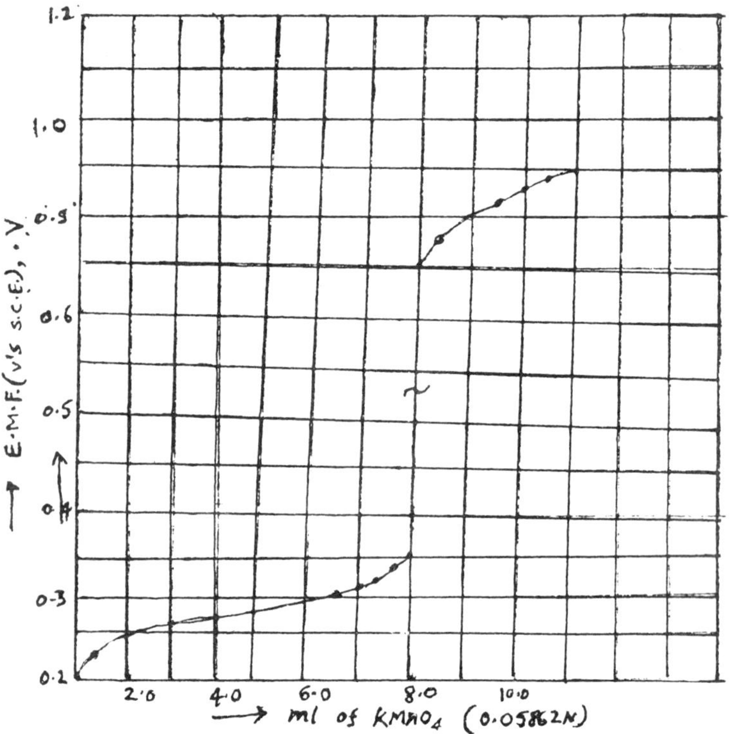

Fig 10.1 Potentiometric titration of 0.05N iyon (ii) sulphate with potassium permangnate.

CHAPTER 11

Precipitation Titrations

Precipitation titrations are those titrations in which the titration reaction results in the formation of a precipitate or a slightly soluble salt. The basic requirements are that the precipitation reaction attains equilibrium very quickly after each addition of titrant; the interferences are absent and the indicator shows a clear end point. Few precipitation reactions are used in such titrations the main reason the lack of suitable indicators to detect end point in precipitation reaction. However, the estimations of halides – Cl^-, Br^-, I^- — using Ag (I) commonly known as argentometric titrations are very old and important methods of analysis. These will be considered in detail.

Three types of argentometric methods are in use depending upon the type of indicators employed.

1. **Mohr's Method** – developed in 1856 by Mohr–this method involves the titration of halides against silver nitrate is a neutral medium, using a 2% solution of potassium chromate as an indicator. Addition of an excess of silver nitrate near the equivalence point result in the formation of a brick red precipitate which marks the end point.

$$Ag^+ + Cl^- \longrightarrow \underset{\text{(white)}}{AgCl \downarrow}$$

$$2\,Ag^+ + CrO_4^{2-} \longrightarrow \underset{\text{(red)}}{Ag_2\,CrO_4 \downarrow}$$

The solution should be neutral but not basic so as to precipitate Ag as $Ag(OH)_2$, with too much acid the end point will be delayed due to lowering of $CrO_4{}^{2-}$ concentration according to the reaction.

$$H^+ + CrO_4{}^{2-} \rightarrow HCrO_4{}^-$$

Under proper conditions Mohr's method is accurate and applicable to low Cl^- concentration.

2. **Volhard's Method.** This method involves the titration of Ag(I) with NH_4SCN using ferric salt as an indicator. Being titration Ag (SCN) is formed. At the end point excess of NH_4SCN reacts with Fe(III) to form deep red $[Fe\ SCN]^{++}$. The amount of thiocyanate which will give a visible colour is very small. Thus the end point error is very small. However, the solution should be shaken vigorously at the end point as Ag ions are absorbed on the precipitate. In Volhard's method, one can easily determine chloride ions in acidic solution (in basic media Fe^{3+} will hydrolyse). An excess of $AgNO_3$ is added to chloride solution and the excess is back titrated with NH_4SCN using ferric alum as in indicator.

AgSCN is less to soluble than AgCl; this results in the reaction :

$$AgCl + SCN^- \rightarrow AgSCN + Cl^-$$

This will consume more NH_4SCN and the chloride content will appear lower. This error can be eliminated by filtering off AgCl before back titration. If little nitrobenzene is added, it will adhere to AgCl and protect it from reaction with thiocyanate but nitrobenzene slows down the reaction.

3. **Fajan's method.** Fajan (1923-24) employed adsorption indicators for the detection of end point in the estimation of Ag^+ ions (or Cl^- ion) against chloride solution (or $AgNO_3$) solution). The following indicators are usually employed for the purpose :

(*i*) Fluorescein : (Used in the determination of Cl^- ions in a neutral or weak acidic solution).

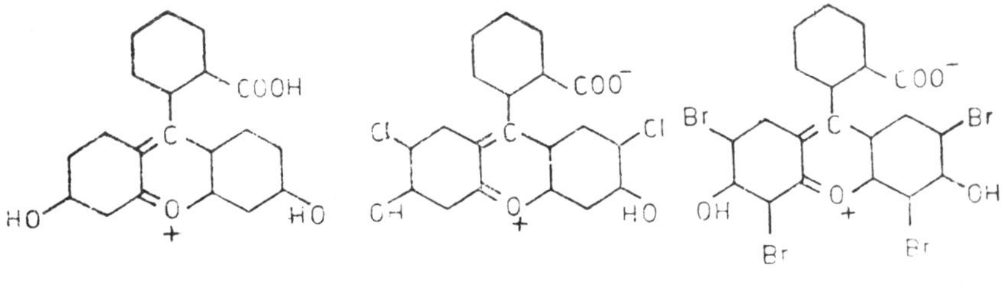

Fluorescein(I), Dichloroflurescein(II), Eosin(III)

Fig. 11.1. Fluorescein(I), Dichloroflurescein(II), and Eosin(III)

(*ii*) Eosine : (Used in the determination of bromide and iodide ions in a weak acidic solution).

(*iii*) Dichlorofluorecein : (Used in the determination of Cl^- ions in a weak acidic medium).

(*iv*) Di iododiethylfluorescein : (Used in the determination of I^- ions to a neutral or weak acidic medium).

Working principle of Adsorption indicators. During the course of titration, the ions present in excess are adsorbed on the surface of the precipitate forming a primary large. For example in the titration of Cl^- solution against $AgNO_3$, silver chloride is precipitated. Silver chloride adsorbs Cl^- ions (present in excess). This is called a primary adsorbed layer. The chloride ions present in this layer generally adsorb oppositely charged ions (say Na^+ or K^+ ions from NaCl or KCl) present in solution. This layer is called a secondary layer. As the end point is approached Ag^+ ions becomes in excess. Now Ag^+ ions are preferentially adsorbed by AgCl forming a primary adsorbed layer. And in term this layer will absorb NO_3^- ions.

At the end point the negative fluorescein ions are adsorbed in preference to NO_3^- ions forming a pink coloured complex with Ag^+ ions.

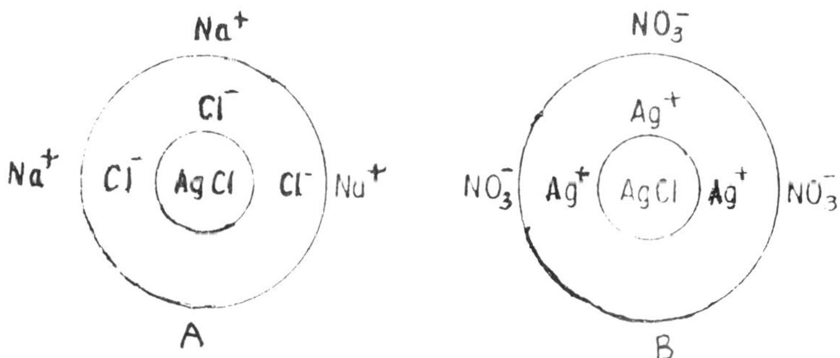

Fig. 11.2. (*a*) AgCl precipitated in the presence of excess of Cl^- ions.

(*b*) AgCl precipitated in the presence of excess of Ag^+ ions.

Experiment. Determine the strength of $AgNO_3$ solution using K_2CrO_4 as the indicator (Mohr's method). Prepare a standard solution of sodium chloride.

Procedure. Prepare a standard 0.05 NaCl solution (0.2925 g for 100 ml solution) by weighing the requisite amount, in distilled water.

Rinse and fill the burette with a given $AgNO_3$ solution. Take 10 ml of N/20 NaCl solution in a titration flask. Dilute the solution by adding 20 ml of Cl^- free distilled water. Now add 5 drops of 2% K_2CrO_1 solution. Add silver nitrate solution gradually, stirring at each addition, till a light brick red colour appears. This is the end point. Note down the burette reading. Repeat the procedure to get three concordent readings.

Calculation. Suppose V ml of $AgNO_2 \equiv$ 10 ml of 0.05 N NaCl solution.

or $$N_{AgNO_3} = \frac{10 \times 0.05}{V}$$

$$\text{Strength of } AgNO_3 \text{ solution} = \frac{10 \times 0.05}{V} \times 169.9 \text{ g/1}$$

Precaution. Store $AgNO_3$ solution in a coloured bottle.

Experiment. 2. Determine the amount of NaCl present in a given solution by titration against $\frac{N}{40}$ $AgNo_3$ solution using fluorescein as indicator (Fajan's method).

Procedure. Take 10 ml of the given sodium chloride solution in a conical flask and dilute it with 20 ml of Cl^- free distilled water. Add 5 drops of fluorscein (0.2% solution fluorscein in alcohol) and titrate it with $AgNO_3$ solution while shaking the flask constantly until the solution acquires pink colour (at the outset of titration the colour of the solution is greenish yellow). Repeat the titration to get three concordant readings.

Calculations Suppose 10 ml of NaCl solution

$$\equiv V \text{ ml of } \frac{N}{40} AgNO_3 \text{ solution}$$

$$N_{NaCl} = \frac{V}{40 \times 10}$$

$$\text{Strength of NaCl salts solution} = \frac{V}{40 \times 10} \times 58.5 \text{ g/1}$$

CHAPTER 12

Complexometric Titrations

Complexometric titrations represent an important advancement in titrimetric procedures. Their development is due to the pioncering work of Schwarzenbach and Flaschka. In 1946, Schwarzenbach introduced EDTA (ethylenediaminetetracetic acid) as a titrant and found some metal sensitive indicators from detracting the end point visually. Thus an important new branch of titrimetry involving chelate formation was originated. This is also called 'complexometry', 'chelatometry' and 'chelometry'.

IUPAC has defined 'complexometric titration' as a titration based on reaction of a metal ion with a ligand to form a soluble complex and in which one of the two reactants is used as titrant. If the titration involves the formation of a soluble, chelate, it is known as a 'chelatometric titration' and when a 1 : 1 soluble chelate is formed it is known as 'chelometric titration'. Thus, the titrations involving the use of polyaminopolycarboxylic acid titrants can be covered under all these three terms.

Schwarzenbach introduced the titrations but important contribution in the field were made by the numerous publications of Flaschka, Pribil, Bermejo-Martinez and Wehber so that at present complexometric titrimetric procedures for majority of elements in diverse types of materials are available to the analyst. Moreover, numerous complexometric procedures for the determination of various organic compounds have also been published.

Flaschka has enumerated the following essential requirements for complexation reactions used in titrimetric procedures :

(*i*) The complexation reaction must be stoichiometric so that a basis of calculation exists.

(*ii*) The rate of reaction must be sufficiently fast.

(*iii*) The stability of the complexes must be sufficiently high, otherwise a sharp end point cannot be obtained because of dissociation.

(*iv*) The complex reaction should involve as few steps as possible so that a sharp end point is assured.

(*v*) A simple method for the location of the end point must be known.

(*vi*) No precipitation should occur during the titration to avoid such compilations as coprecipitation, adsorption and other phenomens often accompanying precipitation.

Most of the usual complexing agents reacts slowly and non-stoichiometrically and therefore cannot be used as titrants. Polyaminopolycarboxylic acids, fulfil all the above mentioned conditions and are extensively used as titrants.

Complexing Agents

EDTA (Ethylene diaminetetraacete acid) is the most common complexing agent : It is also known by several other names such as Versene, Complexes III; Sequesters; Nullapon, Trilon B, Idranat III, etc the

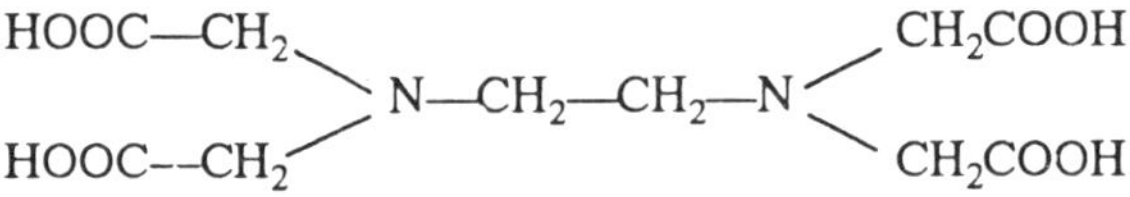

$$\begin{matrix} HOOC{-}CH_2 \\ HOOC{-}CH_2 \end{matrix} \Big> N{-}CH_2{-}CH_2{-}N \Big< \begin{matrix} CH_2COOH \\ CH_2COOH \end{matrix}$$

Structure of EDTA

structure of EDTA has both oxygen atoms and nitrogen atoms as donors and it usually forms six membered chelate rings. Another agent is Nitrilotriacetic acid N $(CH_2COOH)_3$. The calcium complex of NTA is shown in Fig. 12.1

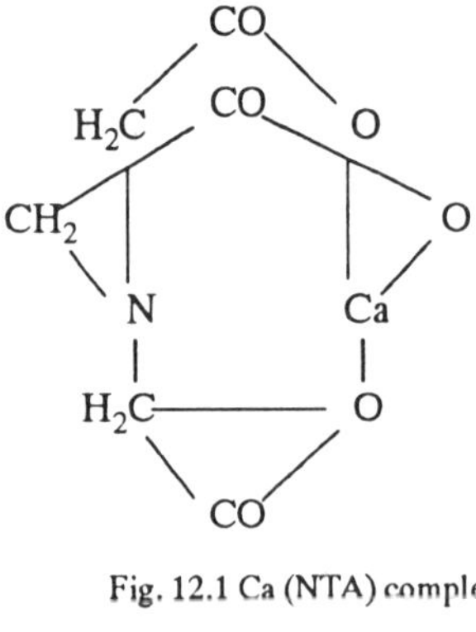

CO
H_2C CO O
CH_2 O
N Ca
H_2C — O
CO

Fig. 12.1 Ca (NTA) complex.

Other complexing agents are EGTA and CDTA (Figs 12.2 and 12.3)

$CH_2—O—CH_2—CH_2—N(CH_2COOH)_2$
|
$CH_2—O—CH_2—CH_2—N(CH_2COOH)_2$

1, 2–d$_i$ (2-aminoethoxylethane)

Fig. 12.2. N N N′ N′ Tetraacetic acid (EGTA)

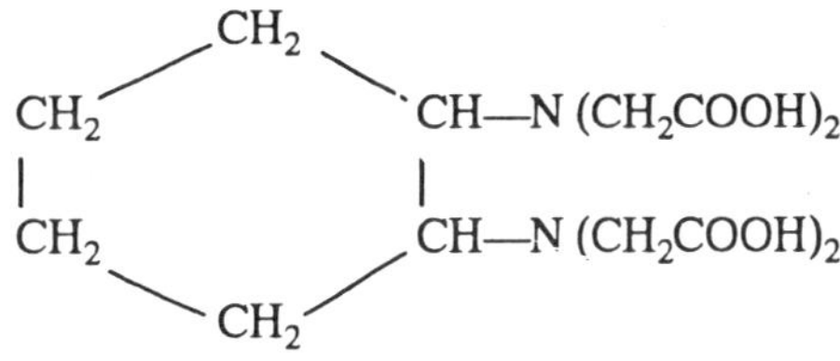

1, 2 diaminocyclohexane

Fig. 12.3 tetraacetic acid (CDTA)

Triethylentetramine (trien) was also used in titrations. EDTA has the following advantages and is the most commonly used :

1. It dissolves readily in water.
2. It is obtainable in high degree of parity.
3. Dihydrate can be used as the primary standard however due to uncertain amount of water it is better to standardise it.

EDTA may be assigned the formula H_4Y; the disodium salt is therefore Na_2H_2Y and is solution yield the ion H_2Y^{2-}. The reaction with cation such as M^{2+} may be represents as :

$$M^{2+} + H_2Y^{2-} \rightleftharpoons MY^{2-} + 2H^+ \quad (1)$$

For trivalent ions and others, we may write

$$M^{3+} + H_2Y^{2-} \rightleftharpoons MY^{-} + 2H^+ \quad (2)$$

$$M^{n+} + H_2Y^{2-} \rightleftharpoons (MY)^{n-4} + 2H^+ \quad (3)$$

One gm ion of the complexing ion H_2Y^{2-} reacts in all cases with one gm ion of the metal, and in each case, also, to two gm ions of H^+ are formed.

From equation (3) it is clear that the ionization of the complex is pH dependent lowering the pH (*i.e.* increasing H^+ ion concentration) decreases the stability of the complex. In general the complexes of trivalent metal ions are stable in basic solutions and that of trivalent and tetravalent ions in acidic medium (pH 1—3).

Metallochromic Indicators

In EDTA titrations a metal-ion sensitive indicator is often employed to detect changes in metal ion concentration $[M^{n+}]$. In the titration curves of EDTA, free metal ion concentration : $pM = -\log [M^{n+}]$ is plotted against the volume of EDTA solution added. A point of inflexion occurs at the equivalence point. The general shape of the titration curves is shown in Fig 12.4

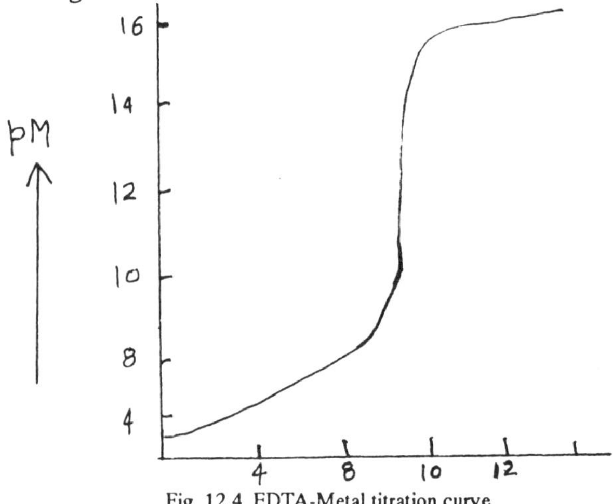

Fig. 12.4 EDTA-Metal titration curve

Metalbochromic indicators are complexing agents themselves. Most commonly used indicators are Eriochrome black T and mureoxide The usefulness of indicator depends upon the stability. The indicators should be half complexed and half free at pM value of inflection in curve.

Eriochorme Black T. This substance is sodium 1-(1-hydroxy 2-naphylazo)-6-nitro-2-naphtal-4-sulphonate. This is also known as Solochrome Black T. In the pH range 7-11 the addition of metallic salts produces a brilliant change in colour from blue to red.

$$\underset{\text{}}{M^{2+}} + \underset{\text{(blue)}}{HD^{2-}} \rightarrow \underset{\text{(red)}}{MD^{-}} + H^{+}$$

This colour change can be observed with the ions of Mg, Mn, Zn, Cd, Hg, Pb, Cu, Al, Fe, TiCo, Ni and the Pt metals. In order to maintain pH constant, a buffer mixture is used.

Eriochrome Black T has the structure :

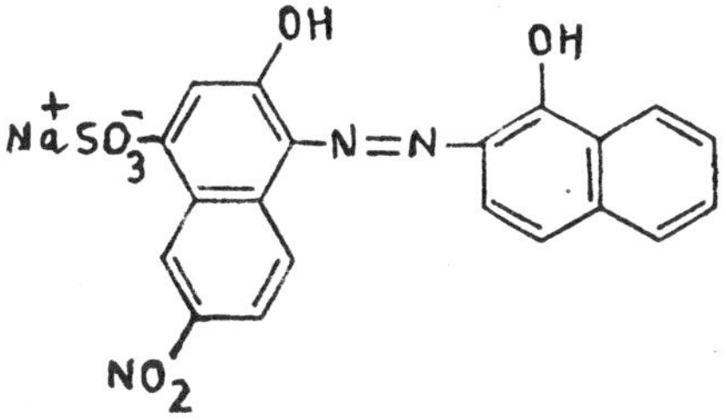

Murexide. It was probably the first metal ion indicator in EDTA titrations. It is the ammonium salt of purpuric acid and has the following structure.

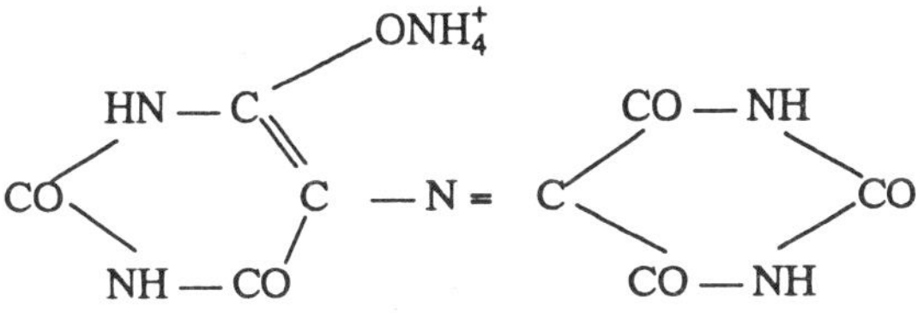

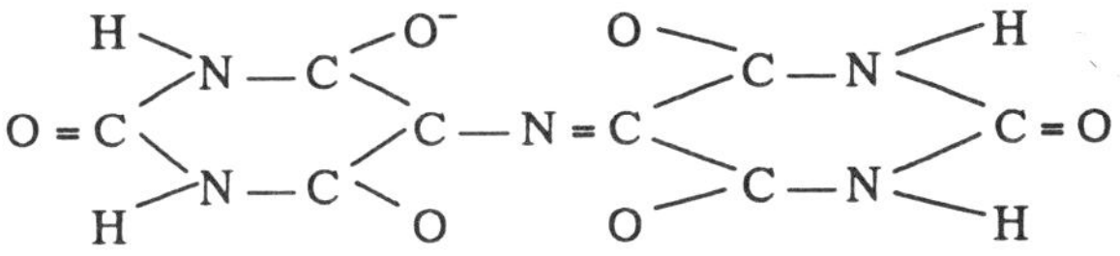

A solution of Murexide is red-violet below PH 9, violet from pH 9 to 11 and blue above pH 11.

Murexide forms complexes with Ca, Ni, Co, Cu, etc., which are sufficiently stable to find application in titration.

Aqueous solution of murexide is not stable for long time, so freshly prepared solution should be employed every day.

Complexometric titrations are classified into 4 types :

(*i*) Direct titrations.

(*ii*) Back titrations.

(*iii*) Replacement titrations.

(*iv*) Indirect titrations.

Direct titration is the simplest and most preferable approach. The sample solution is adjusted to required condition by adding a buffer and an indicator and a standard solution of titrant is added until the colour change indicates the end point. However, direct titration cannot work when (*a*) the metal ion is not kept in solution, (*b*) an indicator is not available, (*c*) rate of complexation of metal of EDTA is very slow, and (*d*) metal ion may not form a suitable stable complex.

Complexmetric titrations are highly versatile procedures. This versatility is an same regent can be used for the determination of numerous ions. The disadvantage is due to the high reactivity of the reagent, the changes of interference by other elements increase and this versatility of complexometric titrations also corresponds to low selectivity. Selectivity can be imported by (*a*) control of pH, (*b*) use of masking and demasking agents, and (*c*) employing the other titrants.

Control of pH. When the stability of metal complexes is sufficiently different, it is possible to carry on selective titration of the metal forming a less stable complex. For example, iron(III) can be titrated with EDTA at pH I without interference from most of the bivalent ions.

Masking and demasking. Masking agents are quite useful to increase the selectivity of a complexometric titration. However, extensive trials should be initially undertaken to compare the efficiency of various available masking agents and to investigate the influence of pH and temperature on the efficiency of masking agents. Sometimes, after titration it is necessary to re-obtain the masked species, necessitating the use of demasking agents.

Experiment 1. Determination of magnesium using Eriochrome Black T as indicator.

Materials required

(*i*) *0.05 M EDTA solution.* Analytical reagent grade disodium salt of EDTA is available commercially and can be used as a primary standard. However, many commercial samples are available which contain a trace a moisture. In such cases, these samples are dried at 80°C and then weighted out directly.

The 0.05 M EDTA solution is prepared by dissolving 18.613 g of A.R. EDTA (mol, wt. = 372.55) in water and diluting to 1 litre in a volumetric flask with redistilled or deionized water.

(*ii*) *Eriochrome Black T indicator.* It is prepared by dissolving 0.2 g of the dye stuff in 15 ml of triethanolamine and 5 ml of absolute ethanol,

(*iii*) *Buffer solution of pH 10.* It is prepared by adding 142 ml of concentrated ammonia solution (sp. gravity 0.88-0-90) to 17.5 g of A.R. ammonium chloride and then diluting to 150 ml with redistilled or deionized water.

(*iv*) *Magnesium ion solution, 0.05.* It is prepared by dissolving about 3.05 g (accurately weighed) or A.R. magnesium sulphate heptahydrate in water and then diluting to 250 ml in a volumetric flask.

Procedure. Pipette out 225 ml of the magnesium ion solution in a titration flask. To this add 75 ml of distilled water, 2 ml of buffer solution (pH = 10) and 3-4 drops of Erio T indicator titrate with 0.05 EDTA solution until the colour changes from red to blue at the end point. Titration should be conducted slowly near the end point.

1 ml 0.05 M EDTA ≡ 1.216 mg of Mg.

Experiment. 2 Determiniation of calcium using Murexide as indicator.

Materials required.

(*i*) *0.01 M EDTA solution.* Prepare a subscribed earlier.

(*ii*) *Murexide indicator solution.* Prepare saturated aqueous solution.

(*iii*) *NaOH solution.* Prepare I *M* aqueous solution.

(*iv*) *Calcium ion solution (0.01* M). It is prepared by weighing accurately 0.50 g A.R. $CaCO_2$ and then transferring the salt in a 500 ml measuring flask. It is dissolved in minimum quantity of 6 MHCl, adding drop by drop until effervescence ; stops and the solution becomes clear. Finally, the solution is made upto the mark.

Procedure. Take 25 ml of the calcium solution dilute to 100 ml with water and add 2 ml of 1 M NaOH solution to have a solution of pH about 12. Add few drops of murexide indicator and titrate against standard EDTA solution. The colour change at the end point is from red to purple.

Calculations. 1 ml of 0.01 *M* Edta ≡0.4001 mg Ca

Experiment 3. Determination of Zine against **Edta**

Materials required

EDTA solution (0.1 M), indicator (EBT) and buffer solution are prepared by method given earlier.

Zinc ion solution. Weigh 1.63 g of pellets and dissolve in HCl. Neutralize the solution with 1 *M* NaOH and transfer it in 250 ml measuring flask. Make up the solution with distilled water. Zinc oxide or zinc sulphate can also be used for preparing the solution.

Procedure. Pipette out 25 ml of the zinc solution in a conical flask and dilute to 100 ml with distilled water. Add 2 ml of the buffer solution (pH 10) and 2-3 drops of EBT indicator. Titrate against 0.1 *M* EDTA solution until the colour until the colour changes from wine red to blue.

Calculation. The amount of zinc ion in the given solution is calculated by the relation.

1 ml of 0.1 *M* EDTA A ≡6.538 g of Zn

Experiment 4. Determination of Ca^{2+} and Mg^{+2} ions in a given solution and estimation of total hardness of water.

Theory. The mixture having Ca^{2+} and Mg^{2+} ion is titrated in one aliquot of the sample solution and the Ca^{2+} ions are titrated in a further aliquot in strong alkaline medium in which Mg^{2+} ions get precipitated. Mg^{2+} ions are then evaluated from the difference.

Materials required. 0.01 *M* Ca^{2+} and Mg^{2+} ions solutions, 0.01 *M* EDTA, 2 M NaOH, buffer pH 10, Murexide and Erio T indicators.

Procedure. (*i*) Pipette out 25 of the solution (Ca^{2+} and Mg^{2}+ ions) into a comical flask. Add 5 ml of buffer (pH = 10) and dilute to about 50 ml with redistilled water. Now add 3-4 drops of Erio T indicator. Warm the solution upto 60°C and titrate with EDTA until the colour of the solution changes from wine red to clear blue. Titrate slowly near the end point. Suppose V ml is the volume (in ml) of EDTA used.

(*ii*) Pipette out 25 ml of the solution (Ca^{2+} and Mg^{2+} ions) into a conical flask. To this add 5 ml of 2 M NaOH solution and 2-3 drops of Murexide indicator. After diluting the solution to about 50 ml with distilled water, titrate with EDTA until the colour changes from red to purple. Suppose V_1 be the volume (in ml) of EDTA used.

Calculations. $V_1 \times 0.4008 \equiv$ of Ca

$$(V - V_1) - 0.2431 \equiv \text{of Mg}$$

In order to obtain the total amount of these ions present in solution, the numbers are to be multiplied by 100/ Q, where Q is the volume (in ml) of the aliquot, *i.e.*, 25 ml.

Estimation of the hardness of water. It is expressed in parts per million, ppm, of Ca.

$$\frac{B}{x} \times 0.4008 \times M = \text{ppm Ca}$$

where B = ml of EDTA

x = ml of sample solution

M = molarity of EDTA

CHAPTER 13

Chromatography

General Principles

The term **chromatography** refers to any separation technique in which the components of a test sample are caused to pass through a column at different rates of speed. In every chromatographic separation, there is a **stationary phase**, which consists of the packing within the column, and a **mobile phase**, which is caused to travel through the column, the test sample is introduced at one end of the column. As the mobile phase passes through the column, each component of the test sample is continuously partiotioned, on distributed, between the phases. The process is similar in principle to a multistage extraction process with a large number of stages.

Chromatographic processes may be classified on the basis of the physical states of the two phases. The stationary phase may be either liquid or solid and the mobile phase either gas or liquid. Thus the processes may be classified, naming the physical state of the mobile phase first, as **liquid-liquid chromatography, liquid-solid chromatography, gas-liquid chromatography, and gas-solid chromatography**. It is also common to refer to the first two types as **liquid-phase chromatography** and to the last two types as **gas-phase chromatography**, on simply **gas chromatography**.

Chromatographic processes may alternatively be classified on the basis of the mechanism whereby the components of the test sample are distributed between the two phases, regardless of whether the mobile phase is a liquid or a gas. On this basis, there are three major classes of chromatographic separations: **adsorption chromatography,** in which the stationary phase is able to adsorb solutes reversibly from the mobile phase; **partition chromatography,** in which the solute is partitioned between the two phases much as in a liquid-liquid extraction process; and **ion exchange chromatography,** in which charged ions are literally traded back and forth between the two phases.

The experimental apparatus and procedures vary widely. For liquid-solid chromatography, the column may be an ordinary buret, packed with granulated solid particles or beads which comprise the stationary phase. The precise graduation marks on the buret are not necessary, and a tube slightly larger in diameter is generally preferable. For liquid-liquid chromatography, the column is similar except that the solid particle are coated with the liquid that is to serve as the stationary phase before the mobile phase is passed into the column; the solid particles serve merely as a support for the stationary liquid. In gas chromatography, the column is typically much longer and smaller in diameter than in the apparatus for liquid-phase chromatography. Again, the column generally is initially packed with solid particles, which serve either directly as the stationary phase in gas-solid chromatography or as a support for a stationary liquid phase in gas-liquid chromatography.

It must be emphasized that chromatography is a method of separation and that any separation must be followed by some measurement if a qualitative indentification or quantitative determination is to be made. In considering further the general principles of chromatograph, let us assume that some sort of measurement is commonly made at the exit end of the column. The measurement of almost any physicochemical property of a solute solvent system including thermal and electrical conductivity, radioactivity, colour, and other spectral characteristics, can be made the basis of a detection device for monitoring the progress of a separation. Let us further assume that the measuring device or detector responds to every component or solute of the test sample, but not to the solvent (whether liquid or gas), which is the mobile phase. Thus, the output of the measuring device is zero whenever pure solvent is emerging from the column and greater than zero whenever a component of the test sample is emerging. The sample is inserted into the entrance end of the column at what we will call time zero ($t = 0$). It is quickly held, through adsorption, through partition, or through ion exchange, by the first portion of the stationary phase. Fig. 13.1. A represents this condition, assuming that the test sample contains two solutes and that even a small fraction of the stationary phase has sufficient capacity to retain virtually all of the sample components. From this time on, the liquid or gas which serves as the mobile phase is caused to flow through the column; this liquid or gas may be the same as the solvent for the test sample or it may be a different one. As the flow continues, the two components tend to separate into distinct bands (Fig. 13.1). Each band migrates through the column and eventually is discharged from the column. The typical response of the measuring device at the exit end of the column is represented in Fig. 13.2.

The process of causing the components of the test sample to move through the column, by the continuous flow of the mobile phase, is called **elution.** The solvent gas or liquid chosen for the mobile phase is the **eluent.**

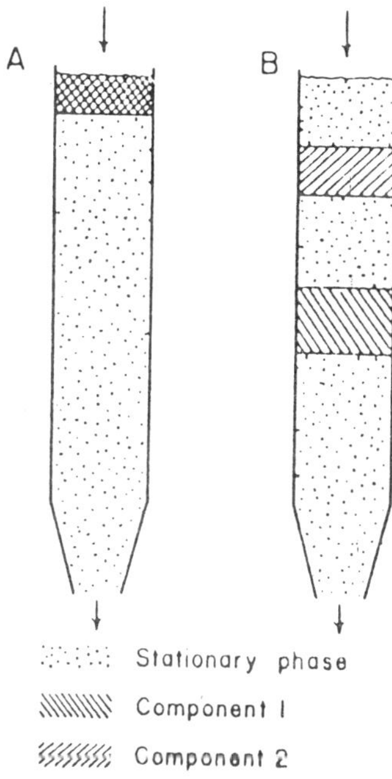

Fig. 13.1. Diagrammatic representation of the retention and elution of a two component mixture on a chromatographic column.

The **eluate** is whatever emerges from the column, that is, the eluate is the eluent plus whatever component of the test sample it contains at any particular time.

The **retention time** for each component is the time required for it to pass through the column, measured from the time of injection or introduction of the test sample to the time of its peak response on the measuring device (Fig. 13.2). The **retention volume** for each component is the volume of mobile phase which must flow to cause that component to pass through the column. If the flow rate is constant, the retention volume is the product of the retention time multiplied by the flow rate.

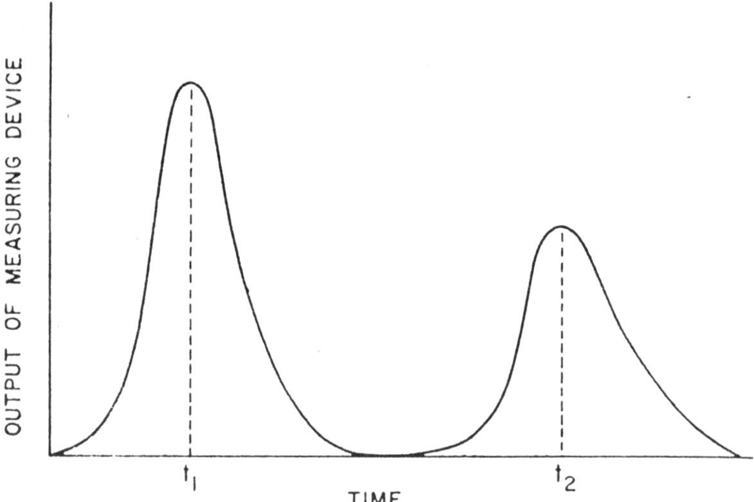

Fig. 13.2. Graphical representation of output of the measuring device (or detector) during a chromatographic separation. The retention time for the first component is t_1, and for the second component it is t_2.

The numerical values of the retention volume and retention time vary from one component to another, which is really the reason why substances may be separated by chromatography. However, these quantities are also influenced by the particular experimental apparatus (column length and diameter, for example) and by the experimental conditions (such as flow rate and temperature). Therefore, the **relative retension volumes** and the **relative retention times** for two or more components within a test sample are more meaningful in describing a chromatographic separation than are the absolute values. The separations often are as distinct and complete as is indicated in Figures 13.1 and 13.2. In many cases, however, there is some everlapping when two or more components are present. Ideally, the solute bands on a chromatographic column should be relatively narrow and should have leading and trailing edges which are perpendicular to the length of the column, as in Figure 13.1. However, a problem of tremendous practical significance to the analytical chemist is **band broadening**. Band broadening may occur in at least two distinct ways. First, it is virtually impossible to pack a column in a perfectly uniform manner; therefore, there is usually a variable now rate in the column due to this irregular packing, and the solute bands send to become uneven and to spread out. Second, any solute in a given solvent has a natural tendency to diffuse under the influence of a concentration gradient; since diffusion is a time-dependent phenomenon, the longer a given solute remains in a chromatographic column before it is eluted, the wider and more uneven will be the corresponding solute band. As a consequence of these effects, much experimental work has been directed toward minimizing band broadening and improving the sharpness of separations in chromatography.

The name **chromatography** is derived from two Greek words meaning "color" and "to write." The earliest applications of this general method of separation did involve colored substances, in which cases it is often possible to see the colored bands on the column as in Fig. 13.1. However, color is only incidental to the separation process, and most chromatographic separations are of colorless materials.

Thin Layer Chromatography

Thin Layer chromatography (TLC) is widely used as a modern analytical methane in chemical analysis and research. It is suitable both for organic and inorganic matter on quantities ranging from the nanograms to the micrograms.

The adsorbers generally used in day to day work are alumina and silica gel with or without binders like starch or plaster of Paris.

There are many ways of applying thin layers of powdered solids of their suspensions to the glass plates. The objective is to obtain a uniform layer throughout the length of the plates. These methods may be classified according to the method of application:

(*a*) **Pouring of Layers.** A measured amount of the slurry is put on a given size plate which is placed on a level surface. The plate is then tipped back and forth to spread the slurry uniformly over the surface.

(*b*) **Dipping.** Two plates held together back to back are dipped in chloroform or chloroform-methanlol sturries of the adsorbent.

(*c*) **Spraying.** In this technique a sprayer used for spraying paint is used for the distribution of the slurry on the glass plate. The slurry has to be diluted further in order to allow the sprape operate.

(*d*) **Spreading.** This is the best and most universally applied method for obtaining uniform thin layers on the glass plates. The apparatus consists of two parts: *the aligning tray* on which the plates are set in a line and the *spreader* which takes up the spreading mixture and applies it uniformly in a thin layer. The procedure is shown in.

Preparation of the plates for TLC. In order to obtain better results the glass should be cleaned thoroughly and completely free from grease. Scrub the plates with a cleaning power (*e.g.* Vim, Biz), or immerse overnight in sulphuric acid-potassium dichromate mixture, or a concentrated solution of sodium carbonate. This is followed by brushing under tap water and rising clean with distilled water. The cleaned plates are left to dry on rack at room temperature in dust free atmosphere.

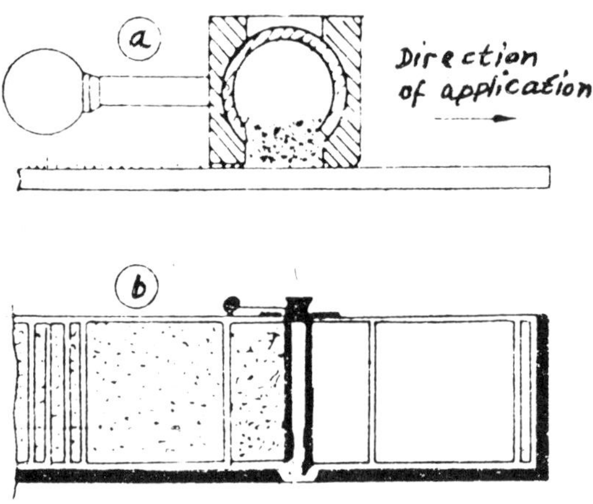

Fig. 13.3 (*a*) Operation of thin-layer spreading (*b*) aligning tray wit glass slide partially coated. (dots=adsorbent).

For coating the cleaned plates are plates in a row, with edges, in alignment. For silica gel plates, 25 of silica gel (silica gel *G*, 200 mesh, containing 13% purified plaster of Paris as binder) is shaken through with 50 ml of distilled water in a 150 ml flask fitted with a plastic stopper for 30 seconds. The thin paste is transferred immediately to the speader which is then pulled across the row of plates with a uniform motion (about 3 seconds per plate), not stopping until all the plates are covered. If wrinkles appear in the thin layers, take each plate while still wet and gently bump the edge of the plate on a hard vertical surface. Clean the spreader immediately, brushing under tap water and rising which distilled water.

The thickness of the layer is usually 0.250 mm, except in preparative work where it may be as large as 2 mm. Leave the prepared TLC plates on a level surface or on the template for about 15 minutes until the surface of the Layer becomes dull, then separate and place in a rack with guide rails and dry them in an oven for $1\frac{1}{4}$ at 110°C for activation. they are then ready for use or they may be stored in a storage cabinet which contains "blue" silica gel as desicent.

For *alumina* plates, 25 g of aluminium oxide *G* is shaken with 50 ml of distilled water and handled as described for silica gel plates. The final activation depends upon the compounds to be separated. 1 or highest activity, alumina plates require 4 hr at 135°C followed by storage over blue silica gel or asumina.

Spotting of the samples on TLC plates. In order to obtain optimum resolution the technique use in spotting is very important. The sample is applied as a solution in an nonpolar a solvent as possible, since the use of a polar solvent has a tendency to cause the starting spot to spread out under also may affect the *R*. value of the compounds. The solvent used for dissolving the sample should be relatively volatic one, so that it may be removed easily from the plate before the development is started. The area of the application should be kept as small as possible, because the smaller the area of application, the sharper will be the resolution. In order to keep the size of the spot small, a series of applications is made by allowing the solvent to evaporate after of application. To assist in this evaporation of the solvent, the plate may be warmed prior to the application or a stream of air/warm air may be directed at the sample spot. The

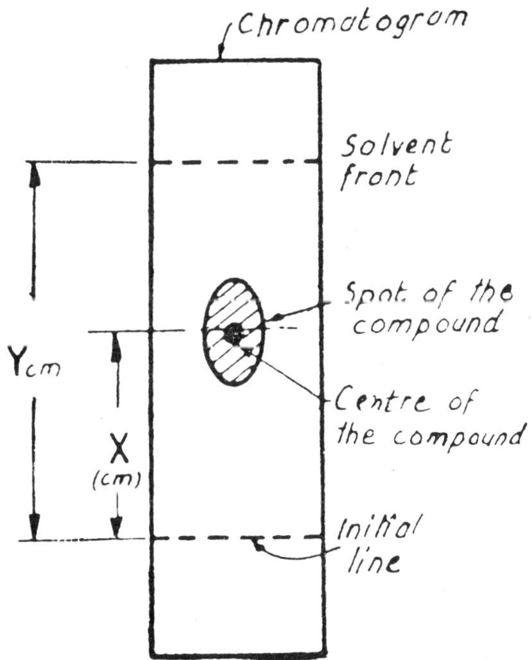

Fig. 13.4 A TLC plate.

samples to be applied are spotted 2 cm from the bottom of the plate at about 1.5–1.0 cm intervals. The samples should be applied when a drawn-out piece of glass tubing having a capillary tip.

Development of the spots is conveniently done in a glass clinder (25 cm in length and 7 cm in diameter) the same as used in paper chromatography.

Visualization. In most of the cases, the substances may not be immediately visible on the plate after the separation. Several methods have been used for visualization. Some of these are :

(1) Spraying of the chromatogram with a reagent in a solvent to bring about a colour change in the substances chromatographed.

(2) Spraying the developed chromatogram after drying with distilled water. As long as the plate is wet the substances show up as white spots against a translucent background.

(3) Visulization under UV-light for compounds which fluoresce. For compounds which absorb in the UV-range. Spraying the chromatograms with solutions of organic fluorescent dyes is done. This makes the UV-absorbing organic substances appear as dark spots on a fluorescent background when viewed in UV-light. But a better way of using the fluorescent indicators is to add them to the adsorbent material before coating. This prevents the difficulties encountered in even spraying which is a must to ensure a high sensitivity of detection.

Some of the fluorescent indicators are : Fluorescein (0.2% solution in ethanol), Morin (0.1% solution in ethanol); Rhodamine *B*, sodium fluoresceinate (0.04% aqueous solution), Zinc silicate, Cadmium silicate, Barium diphenylamine sulphonate, Methyl umbelliferone.

Experiment. 1. *Thin layer chromatographic separation of anthraquinones.*

The purpose of the experiment is to show the separation of closely related compounds. Each one of the anthraquinone dyes has a distinctive colour,. which enabler the spots to be seen visually.

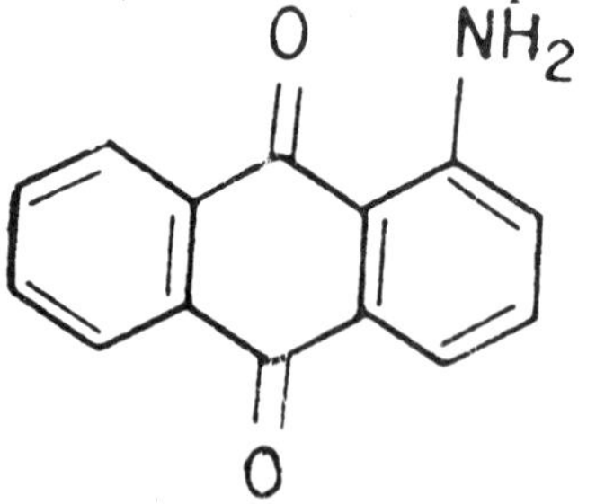

1-aminoanthraquinone

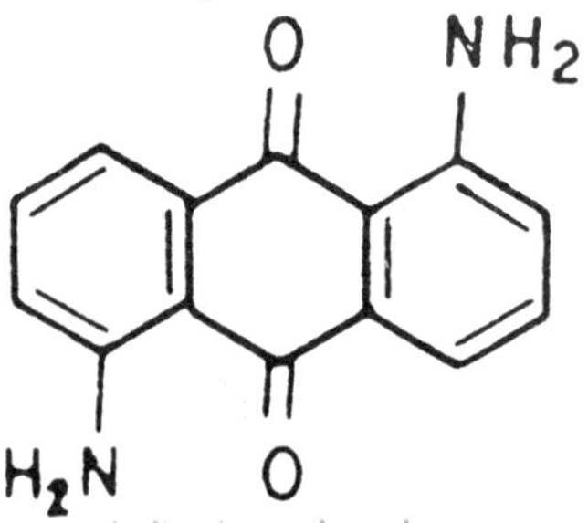

, 5-diamino anthraquinone

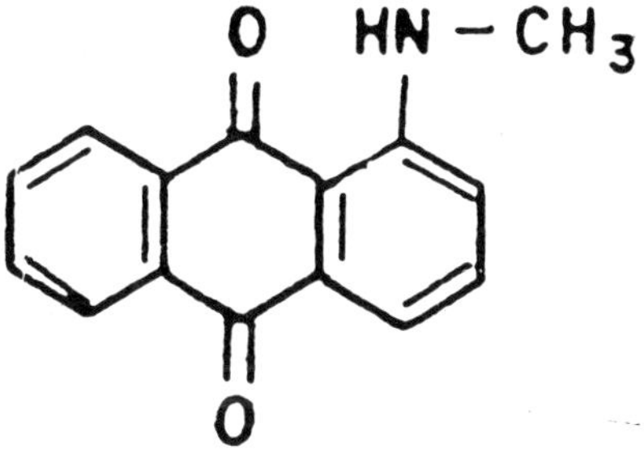

1-methylaminoanthraquinone

The six anthraquinone derivatives which may be used are shown below:

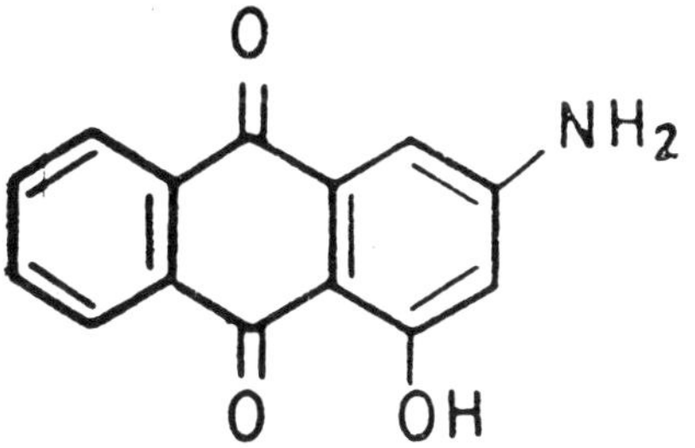

1-amino-4-hydroxy anthraquinone

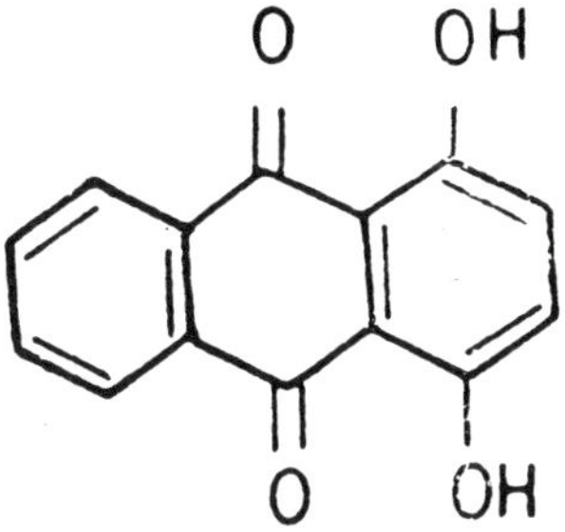

1, 4-dihydroxy anthraquinone

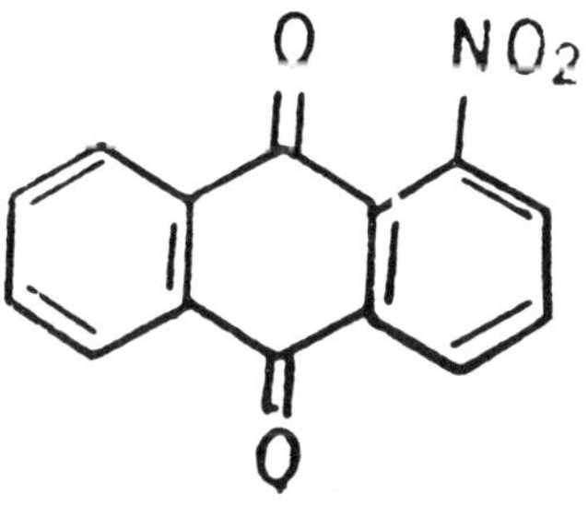

1-nitroanthraquinone

These days have differences in their polarity (i.e., dipole moment is different), so adsorption energies of these molecules will be different.

Solution. Prepare the following solutions:

(*a*) Saturated solution of each of the dyes in ethanol.

(*b*) A saturated solution of all the dyes in ethanol (known mixture) and solution of unknown containing some or all of the dye in ethanol.

Developing solution. 1 : 15 mixture of ether in benzene.

Spotting of plates. Each TLC plate is spotted with three solutions, two spots of standard dyes and one spot of the known or unknown mixture. Each spot is placed 1.5 cm apart from each other. Mark carefully the spots.

Development or chromatogram. Place the spotted plates in developing chamber containing developing solvent. Develop for 10 minutes. Observe the spots while plate is still wet (since the colour, spots are clearest when the strip is still wet). Allow the plates to dry.

Identification of unknowns. Most of these dyes are visible under ordinary light and several of them give fluorescence under UV light. 1-Amino-4(OH) anthraquinone gives blue fluorescence, 1, 4–dihydroxy anthraquinone fluoresces blue green and 1-nitroanthraquinone fluorescences to pale blue. Calculate Rf value and compare them.

Experiment 2. Separation of the components of a commercial analyesic mixture (A.P.C.)

Several commercial remedies for common cold and headache are mixtures of caffeine, salicylamide. Acetyl salicylic acid phenacetin. To check for purity and adulteration of these remedies it is necessary to establish method for the analysis of these preparations.

The Rf values and polarities of these molecules differ from each other and thus these can separated from each other.

Solution. Prepare 1% (w/v) solution of each of the four possible components by dissolving 50 mg in 5 ml of ethanol.

Developing solvent. (*i*) It is mixture of methanol : acetic acid : diethylether : benzene in 1 : 18 : 60 : 120 ratio. This should be freshly prepared.

Spraying reagent. Ferric chloride $K_3[Fe(CN_6)]$ regent. This is prepared by dissolving 0.5g of K_3 [*Fe* $(CN)_6$] and 2.0 g of $FeCl_3$ in 100 ml of distilled water. This should be freshly prepared.

Preparation of sample. In a test tube, powder two tablets of the commercial product with a glass rod. Add 2 ml of methanol and stir for 5 minutes to extract all of the soluble ingredients from the power. Filter the solution through filter paper.

Development of chromatogram. Spot the TLC plate with a known component and with an unknown component as usual and develop the spots in the developing solvent.

Detection of the spots. Salicylamide, acetyl salicylic acid, caffeine and phenacetin can be detected by their fluorescence under UV light at 254 nm. The fluorescence colours are:

Phenacetin............Large uneven dark blue-purple

Salicylamide...........Bright, brilliant blue spot

Acetyl salicylic ....Purplish brown spot
acid

Caffeine ....Dark spot.

Spray the developed chromatogram with $FeC_3l - K_3[Fe(CN)_6]$ reagent if there is some uncertainty in the observation of their fluorescence in **UV light**. The colours produced are:

Acetyl salicyclic acid......brown spot

Phenacetin .... Blue spot.

To detect caffeine, spray the chromatogram with 10% aqueous chloramine-*T* solution and then with 1 N HCl. Heat the sheet or plate in the oven at 9 –98° C, until the Cl_2 odour has vanished. Place the plate in NH_3 vapour, warm slightly the exposed plate–a pink red spot will appear.

Calculate Rf values and compare with the standard value. It is quite probable that salicylic acid may be produced by the hydrolysis of acetyl salicylic acid during the preparation of chromatogram. These will appear as a faint spot around $R_f = 0.85$.

Some commercially available analgesic mixtures containing Acetyl sali-ylic acid, caffeine, phenacetin are listed in the following table :

Name	*Acetyl salicylic acid*	*Caffeine*	*Phenacetin*
ASPRO	√	√	√
A.P.C. Tablets	√	√	√
CODRAL	√	√	√
DRISTAN	√	√	√
APIDIN*	√	√	√
VEGANIN*	√	√	√

* These contain one more constituent in a very small amount.

Experiment 3. Separation of cis⁻ and trans dichlorobis-ethylenediaminecobalt (III) chloride $[Co(en)_2 Cl_2]Cl$.

These *cis* and *trans* isomers may be written as:

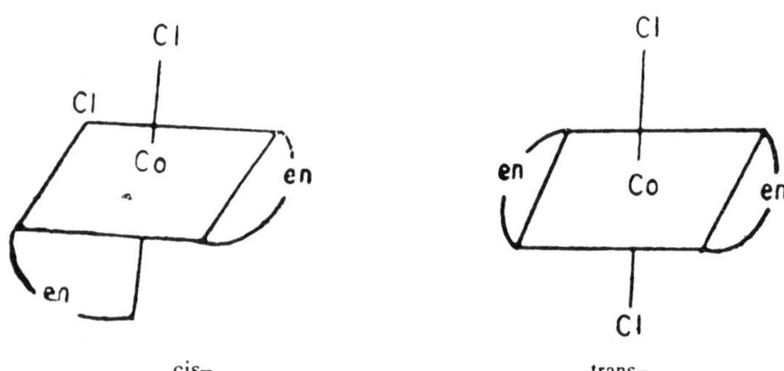

cis– trans–

Though the chemical formula of the compounds is same, yet they differ their polarity (*cis* being more polar than *trans*). Thus *trans* form will have more R_f value.

Preparation of solution. 1% Methanalic solutions of each of the two isomers as well as one of a mixture containing two isomers will have to be prepared. These are prepared by dissolving 10 mg of each isomers in a drop of water and then diluting to 1.0 ml. with methanol. Since the solutions are unstable, they should be prepared a fresh.

Developing solution. It consists of 90 ml methanol, 10.5 ml of 0.5 M sodium acetate in methanol, 0.1 ml of 1 M acetic acid in methanol and 1 ml of water. The solution should be prepared freshly.

Spraying reagents. (*i*) **2 M** aqueous ammonia (*ii*) 0.1% Rubeanic acid in ethanol.

Spotting and development of chromatogram. With the help of the capillary, put the spots of two isomers and one spot of the mixture preferable 1.5 cm apart. After spotting put the plate in developing chamber and allow the solvent to ascend till it reaches to the upper end of the plate.

Detection of spots. Use glass or plastic sprayers. Spray the dried plates with 2 M aqueous ammonia solutions. After allowing the ammonia to dry, spray the plates with 0.1% rubeanic acid in ethanol. Blue-green spots should be produced by each of the isomers.

Identification of unknowns. Calculate the R_t value. and compare with known sample., *Trans isomer* has more R_t value than cir-isomer (*trans* isomer is less polar than *cis* isomer).

Preparation of cobalt isomers in the laboratory. The procedure is described here, so that students can themselves prepare these isomers in the laboratory.

Materials required

Cobalt chloride, $CoCl_2.6H_2O$ = 12 g (0.05 mol)

Ethylene diamine 10% = 60 ml (0.1 mol)

Procedure. To an aqueous solution of $CoCl_2.6H_2O$ (12 g) in distilled water (~ 20 ml) in a beaker (400 ml) add 10% of ethylene diamine (60 ml) with vigorous stirring. Heat to boiling. Transfer the whole solution to a Buchner flask and draw a stream of air by a water jet pump or other arrangement for a period of 10 to 12 hours. Add concentrated HCl (35 ml) and transfer back to beaker. Evaporate the solution, which has turned from dark red to green, on a stream bath until a thick crust is formed. Cool and keep it overnight Bright green-square plates of *trans*–$[Co(en)_2Cl_2]Cl$ are deposited. Filter on a Burner funnel and wash successively with alcohol and ether and dry at 110°C. Yield is ~ 8.9 g, further purification can be affected by dissolving the salt in 25 ml of H_2O and then adding methanol to the filtered solution. On cooling pure form is obtained.

Preparation of cis-form. Place 2 g of trans–$[Co(en)_2 Cl_2]Cl$ in an evaporating dish. Add 25 ml water and immediately evaporate on a steam bath. Scrap the solid from the dish on to an ice-cold sintered glass crucible. Add 12 drops of ice cold water. Turn vacuum after stirring with glass rod. The *cis* form (violet) is left on the filter paper. Wash the violet form with two 10 ml portions of dry ether and dry in an oven below 110°C.

Paper Chromatography

Paper chromatography is an important and useful class of partition chromatography. In this techniques, the stationary phase is considered to be made up of water molecules bond to the cellulose network (inert supper) of the paper. The mobile pase, known as the *developing solvent or irrigant* consists of either one solvent or a mixture of different solvents. Separation of the mixture into pure compounds takes place by the partitioning of different compounds between these two liquid phases (one is water of paper and other is the organic solvent used as irrigant). The mobile phase travels by capilla action through the paper. Depending upon the way the solvent travels on the paper, there are three types of paper chromatography as given below:

(*i*) *Ascending paper chromatography.* The mobile phase moves upon paper strip in this case.

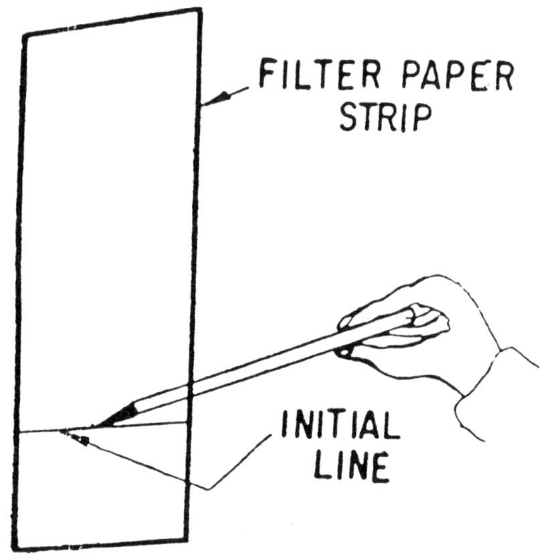

Fig. 13.5

(*ii*) *Descending paper chromatography.* The mobile phase in this case moves down the paper strip.

(*iii*) *Circular paper chromatography.* The mobile phase moves horizontal along a circular sheet of paper here.

The ascending paper chromatography techniques is discussed in detail here.

Experiment 4. *General Procedure of ascending paper chromatography of sugars.*

A 2-3 percent solution of the given sugar (or mixture of sugars is prepared in water in a test tube. This solution is known as the *test solution.* A paper strip of the size, 20×5 cm is out from a Whatman filter paper No. 1 sheet. A horizontal line known as the *initial line,* about 3 cm from one edge of the paper strip is drawn with a pencil. In the middle of this line, the solution is spotted with the help of a very fine capillary as shown in Fig. 13.5 While spotting the solution care should be taken so that the spot does not spread much (not more than 3 mm in diameter). This can be easily checked by just putting the capitallary (containing the solution) on the paper, taking it off quickly and letting the spot dry. This process is repeated 3-4 times to leave a good amount of the sugar on the paper. The spot is known as the *organic.*

A glass cylinder (25 cm in length and 7 cm in diameter) is taken. A rubber cork (which tightly fits into the mouth of the cylinder as shown in

Fig. 13.6. About 15 ml of the developing solvent (butanol : pyridine : water :: 6 : 4 : 3 or isopropanol : water :: 4 : 1) is taken in the cylinder. One end of the paper which is away from the initial line is fixed into the

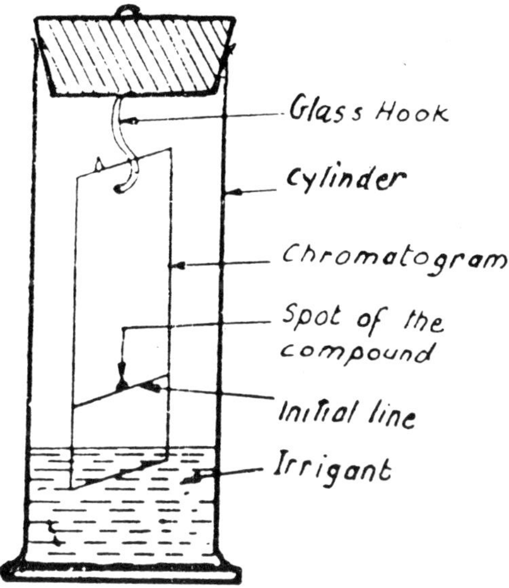

Fig. 13.6

glass hook in its middle and hung as shown in Fig. 13.6. Care should be taken that only 1 cm length dips into the developing solvent. When the solvent has risen about 10-12 cm from the initial line, the paper is taken out and a line known as the *solvent front*, is marked with a pencil at the level to which the solvent has risen. The paper is then dried in air and is known as *chromatogram*.

The sugar of the mixture would be somewhere between the initial line and the solvent front. They are detected by converting them into coloured derivatives. This conversion is brought about by spraying with aniline–diphenylamine reagent (prepared by mixing 10 volumes of a 1% solution of diphenylamine and 1% aniline in acetone with 1 volume of 85% phosphoric acid) with the help of a fine sprayer as shown in Fig. 13.7 and a spraying bulb (Fig. 13.8). The paper strip is heated in an oven at 100-110°C for about 5 minutes. The number of spots on the paper indicates the number of constituents of the mixture. The distances from the initial line to the centre of each spot and that to the solvent front are measured.

The ratio,

$$\frac{\text{distance travelled by the compound (x) cm}}{\text{distance travelled by the solvent (y cm)}}$$

is constant for a compound at a particular temperature and in the same solvent system. This constant is known as the R_f value of the compound (R_f *stands* for *retentionfactor*). It is useful in comparing the compounds.

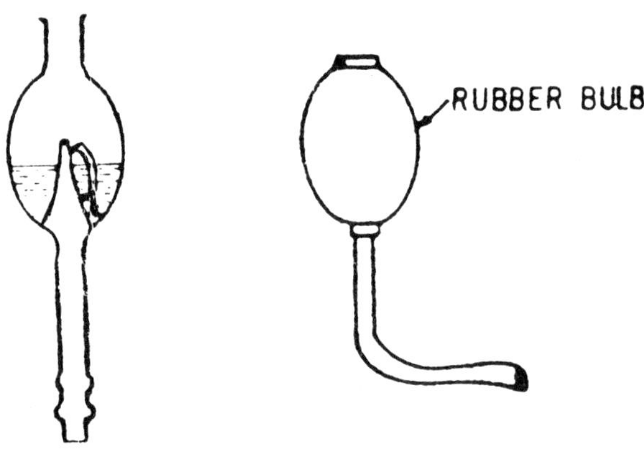

Fig.13.7 A glass sprayer.

Fig. 13.8 A rubber bulb.

Precautions

1. While cutting the paper strip from a Whatman filter paper sheet, care should be taken that the paper is cut lengthwise along the machine end of the sheet. The machine end can easily be found out by putting a small drop of water on the corner of the filter paper sheet. The drop will spread in an elliptical shape. The end along the major axis of the ellipse is the machine end of the paper.
2. The paper strip before use should be saturated with the vapours of the developing solvent to be used to get good results. This can be done by keeping the strip in the jar containing the developing solvent. Care should be taken that the strip drop not dip into the solvent.
3. The marking pencil should be of a good quality lead, i.e., the lead should not dissolve in the irrigant.
4. While spotting the sugar solution on the paper, care should be taken so that the size of the spot does not become too big.
5. The jar should not be disturbed during the period the solvent is rising.

Recording

Spot No.	Distance between the initial line and the solvent from (cm)	*Distance between the initial line and the centre of the developed spot (cm)*	R_f *value*	Tempe-rature (C°)	*Colour of spot*
1					
2					
3					

Result

The given test solution contains.......different sugars.

The following sugar can be used ; the colours they give with the spraying agent (aniline, diphemylarine) are given below:

Glucose........Blue

Sucrose........Brown

Fructose ....Brown

Lactose.........Blue.

Preparation of aniline hydrogen phtalate as spraying reagent for the identification of sugars. Take 100 ml of distilled water and 15 of n-butanol. Shake the mixture well and leave for 12 hours. Take the lower layer and add to this 0.93 g of aniline (approx. 2 ml) and 2 g of phthalic anhydride. Shake well and use this for praying the chromatograms for the identification of sugars.

R_f value for sugars (*n-butanol* : acetic acid : water 4 : 1 : 5)

D Glucose	0.18
D Fructose	0.23
D Ribose	0.33
Lactose	0.09
Sucrose	0.14
L Ascorbic acid	0.38

Experiment 5. *Separation of amino acids by paper chromatography.*

R_f values of amino acids.

	Butanol solvent	*Phenol solvent*		*Butanol solvent*	*Phenol solvent*
Histidine	0.07	0.69	Proline	0.30	0.91
Serine	0.10	0.36	Tyrosine	0.32	0.64
Lysine	0.10	0.48	Methionine	0.40	0.80
Arginine	0.11	0.59	Valine	0.47	0.77
Aspartic	0.13	0.15	Tryptophane	0.47	0.83
Glutamic	0.16	0.25	Isoleucine	0.55	0.86
Glycine	0.17	0.40	Phenylalanine	0.58	0.89
Alanine	0.22	0.34	Leucine	0.60	0.85
Theorine	0.22	0.50			

Phenol solvent. Phenol: water (400 : 100 + 25 mg 8-hydroxyquinoline ; in an atmosphere of HCN, 100 mg NaCN + 5 ml H_2O).

Butanol solvent. n-Butanol : glacial acetic acid: water (4 : 1 : 5).

Procedure. Prepare a mixture of n-butanol, acetic acid and water in the above proportions, shake well ; then allow the phases to separate. Put the lower phase in a cylinder. Hand the spotted paper into the cylinder taking care that only about 1 cm length dips into the developing solvent. Allow the chromatogram to develop for several hours. When the solvent front has risen about 10-12 cm from the initial line, remove the paper from the cylinder, mark the solvent front and allow it to dry in air. When the paper is dry spray the chromatogram with 0.1% ninthydrin in 95% ethanol +5% collidine. Put the papers in the oven at 60°C until the first blue spots are faintly visible. Store the chromatograms in the dark overnight and reexamine for full colour development. Calculate R_f values of the different amino acids.

Application of Paper Chromatography in Qualitative analysis of Some Inorganic Cations.

Experiment 6. *Paper chromatographic separation of Cu(II) and Cd(II).*

Materials required. (*a*) Paper chromatographic jar, Whatman No. 1 filter paper strip, capillary tubes for spotting.

(*b*) Salt solution : Aqueous 1% salt solutions of copper sulphate and cadmium sulphate.

(*c*) Developing solvents.

(*i*) Ethanopl (90 parts) + 5N HCl (10 parts) both by volume).

(*d*) Reagents :

(*i*) Colourless $(NH_4)_2S$ solution prepared by bubbling H_2S through dilute NH_4OH.

(*ii*) 0.5% Dithiazone in chloroform.

Procedure. Put into the jar a depth of 1 cm of a solvent. On a chromatographic filter, paper put the initial line about 1.5 cm from the bottom and along this line place small spots of Cu^{2+} and Cd^{2+} ions with the help of capillary tubes as shown in Fig. 1. Put the spots at least 2 cm apart. Allow the spots to dry. The paper strip is put vertically in the jar containing the solvent as shown in Fig. 2. Put a clock glass on the jar.

Allow the solvent to climb up the paper until the front edge of the solvent is almost near the top of the paper. Remove the paper from the jar. Mark carefully the position of the front edge of the solvent and allow the paper to dry.

Develop the spots by spraying colourless $(NH_4)_2S$ solution or 0.5% dithiazone in chloroform. Dry the paper.

$(NH_2)_2S$ *solution*	*0.5% dithiazone solution*
Cu—Black	Cu—Purple brown
Cd — Yellow	Cd — Purple

Calculate the R_f values of the two ions as explained earlier.

Measure the distances travelled by the solvent front and the respective ions.

Calculate the R_f values for the two ions.

$$R_f \text{ value for the ion} = \frac{\text{Distance travelled by the ion}}{\text{Distance travelled by the solvent front}}$$

Experiment 7. Paper chromatographic separation of Ni(II) *and* Co(II)

(*a*) Items as given under the previous experiment.

(*b*) *Salt solutions.* Aqueous 1% solutions of nickel sulphate and cobalt sulphate.

(*c*) *Development solvent.* Acetone (90 parts) + concentrated HCl (5 parts) + Distilled water (5 parts) – all by volume.

(*d*) *Visualizing reagent.* Fume the chromatographic paper strip over ammonia for about 2 minutes. Then spray the strip with a solution of 0.5% of dithioxamide in ethanol.

(*e*) *Procedure.* As given in the previous experiment

Nickel spot	Bluish purple
Cobalt spot	Yellowish orange

Calculate the R_f values for the ions.

Experiment 8. Paper chromatographic separation of As(III) *and* Sb (III).

Materials required.

(*a*) Items as given under (*a*) in the previous experiment.

(*b*) *Salt solution.* Aqueous 1% solution of antimonous chloride (HCl is added to supress hydrolysis) and arsenic chloride (prepared by dissolving AS_2O_3 is NaOH and neutralizing the solution with dilute HCl).

(*c*) *Developing solvent* :

(*i*) Ethanol (10 parts) + 5 M HCl (90 parts) – All by volume

(*ii*) Ethanol (45 parts) + iso-propanol (45 parts) + 5 N HCl (10 parts) – All by volume

(*d*) *Visualizing reagent.* As given in the previous experiment.

Arsenic spot	Yellow
Antimony spot	Orange

Calculate the R_f values for the two ions.

Experiment 9. Paper chromatographic separation of Fe(III) *and* AI(III).

Materials required.

(*a*) Items as given under (*a*) in the previous experiment.

(*b*) *Salt solution.* Aqueous 1% solution of $FeCl_3$ and $AlCl_3$ (HCl is added to avoid hydrolysis).

(*c*) *Developing solvent.* Ethanol (45 parts) + Iso-propanol (45 parts) + 5 M HCl (10 parts) – All by volume

(*d*) *Visualizing reagent.* 1% Alizarin solution in ethanol.

(*e*) *Procedure.* As given in the previous experiment.

After spraying the chromatographic strip with 1% alizarin solution, expose the strip to ammonia vapour and finally warm the filter paper.

Aluminium spot	Red
Iron spot	Purple

Calculate the R_t values for two ions.

CHAPTER 14

Electroanalytical Techniques

Analytical methods based on the electrochemical properties of solution are termed as electroanalytical techniques. An electrode when immersed in an electrolyte solution constitute a half-cell. When two half-cells are combined and connected externally, the cell will act as a source of electrical energy and produce current.

We can define half cell potential as

$$E_A = E_A0 - \frac{RT}{nF} In \frac{[A_{red}]}{[A_{ox}]}$$

(Where A and B refer to two half-cells.)

$$E_B = E_B0 - \frac{RT}{nF} In \frac{[B_{red}]}{[B_{ox}]}$$

we have the actual value of potential of cell as

$$E_{cell} = (E_B - E_A)$$

In the above expression E_B^0 and E_B^0 are known as standard potentials, for the half reaction. It is difficult to determine the potential of a single electrode as a second electrode is required for measurement. The normal hydrogen electrode (NHE) is usually assigned potential of zero volt. In NHE, H_2 gas at a pressure of one atmosphere is bubbled over a platinised platinum foil immersed in aqueous solution, in which the activity of hydrogen is taken as unity.

The two electrodes of a cell are separated by maintaining electrolytic contact between them. This contact may be made us through a medium of salt (kCl) bridge.

For metals, which are more powerful reducing agents than Hydrogen E° values are negative with respect to NHE. E° value are positive for those which are less powerful reducing agents.

$$Zn^{2+} + 2e^- = Zn \quad E^\circ = -0.76 \text{ V}$$

$$2H^+ + 2e = H_2 \quad E^\circ = 0 \text{ V}$$

$$Cu^{2+} + 2e^- = Cu \quad E^\circ = +0.34 \text{ V}$$

where E° are standard electrode potentials.

Common Electroanalytical methods are:

1. *Potentiometry:* This method is a direct application of Nernst equation through measurement of potentials of two nonpolarized electrodes under conditions of zero current.
2. *Voltammetry and Polarography*: These methods involve the study of the composition of dilute solution by plotting current-voltage curves. The concentration of reducible or oxidisable species at the electrode can be determined. Negative voltage is applied to a small polarizable electrode (relative to the reference electrode) and voltage is increased over a span of 1-2 V and the resulting change in current through the solution is recorded. Voltammetry is refers general name, while polarography refers to applications of the dropping mercury electrode.
3. *Coulometry:* This method of analysis involves the application of Faraday's laws of electrolysis relating the equivalence between quantity of electricity passed and quantity of chemical change occurred.
4. *Conductimetry:* This method employs two identical inert electrodes and the conductance (i.e. reciprocal of resistance) between them is measured. Some methods are dealt with in detail.

POTENTIOMETRIC TITRATIONS

Principles of Potentiometric Titrations

In many cases it is possible to make the solution to be titrated the electrolyte of a half-cell, the potential of which is a function of the concentration of one or more of the species participating in the titration reaction. A wire or foil of an appropriate metal is immersed in the solution to act as the "indicator" electrode. The half-cell thus formed is connected via a salt bridge with a reference electrode. The cell established in this way is connected with a potentiometer circuit and during the subsequent titration, the cell voltage is measured as a function of titrant additions. To determine the end point graphically, the titration curve may be obtained by plotting the values of the half-cell potential ($= E_{cell} - E_{reference}$) versus the volume of titrant solution added. However, because the potential of the reference electrode remains constant, it is common practice to simply plot the total cell voltage rather than the half-cell potential, thus merely shifting the curve vertically without changing its shape or its position relative to the abscissa. A titration so conducted is described as proceeding to a potentiometric end point or simply as a potentiometric titration. Many types of titrations can be effected potentiometrically.

Potentiometric Redox Titrations

A redox titration can be conducted potentiometrically. In this case the indicator electrode is frequently a platinum wire or foil. The point of maximum slope in the curve is taken as the point.

Potentiometric Acid-Base Titrations

The titration curves for acid-base titrations in aqueous solution are plots of pH versus the volume of acid or base added. Such curves are realized experimentally in acid base titrations to a potentiometric end point. An electrode sensitive to hydrogen ion is used as the indicator electrode. A pH meter with a glass electrode can be employed to follow the pH changes during the titration. The measured pH values can be plotted versus the volume of titrant solution added and the point of maximum slope be taken as the end point. Alternatively, if the pH of the equivalence point is accurately known, the titration may be performed without plotting and terminated when this pH value is reached. Potentiometric acid-base titrations may be employed when visual indication becomes difficult due to colorants present in the sample solution. A potentiometric titration is of special value in the determination of acids and bases that are too weak to afford a satisfactory visual titration.

Other Potentiometric Titrations

The presence of an appropriate complexing agent may markedly change the potential of a redox couple involving a metal ion ; consequently, end the point of some compleximetric titrations may be detected potentiometrically. Similarly, the formation of a precipitate by a metal ion that is involved in a redox equilibrium will shift the redox potential. Hence, potentiometric precipitation titrations are also feasible. In the titration of a halide with silver nitrate, a silver wire can serve as the indicator electrode.

Potentiometric titrations often allow the stepwise titration of two or more substances with a single titrant. For example, a mixture of bromide and chloride ion can be titrated with silver nitrate. In the visual titration of this mixture by the Mohr method, the result corresponds to the total of the two halides. In contrast, in the potentiometric titration, two breaks will be obtained corresponding to bromide and chloride, respectively. The accuracy is not especially high because of the closeness of the values of the respective solubility products. More accurate results are obtained in the stepwise potentiometric titration of iodide and chloride. Under favorable concentration conditions, for example, in the analysis of mineral waters, it is even possible to analyze a mixture of iodide, bromide, and chloride by titration with silver ion. Three breaks in the potentiometric titration curve are obtained.

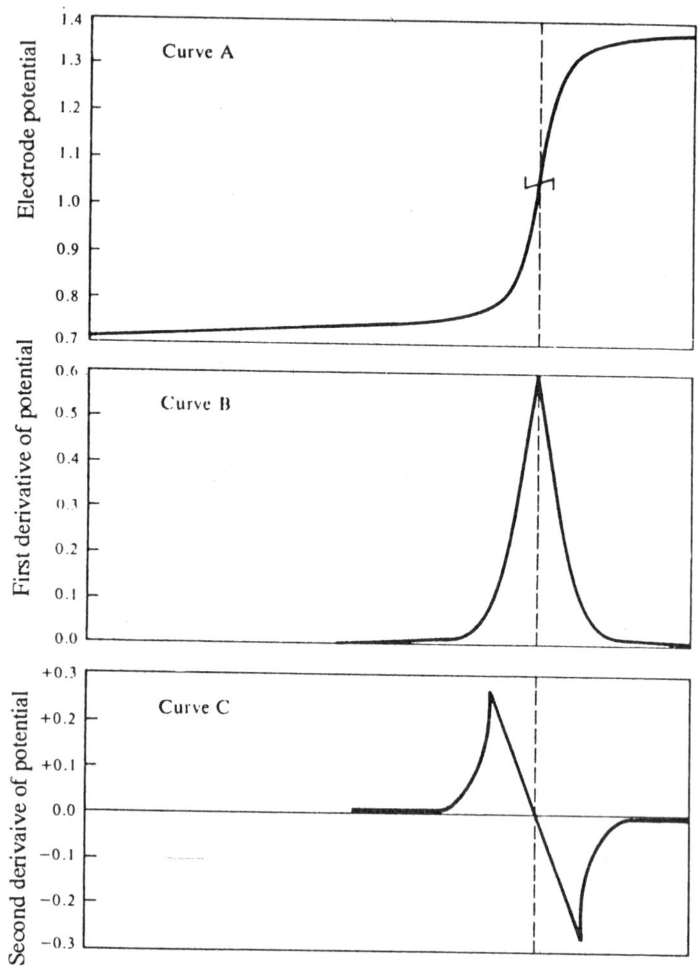

Fig. 14.1. Curve A: experimental potentiometric titration curve
Curve B: first-derivative curve of A.
Curve C: second-derivative curve of A.

Diverse Techniques

Instead of using the inflection point as the end point in a potentiometric titration, more precise results can be gained from the plot of the first or second derivative of the curve. The first derivative curve is readily obtained by adding the titrant solution in small, equal increments and plotting the difference between two consecutive potential (or voltage) readings versus the volume of titrant solution added. The maximum in this curve is taken as the end point. The curve for the second derivative is obtained by plotting the difference between two consecutive differences versus the volume of titrant solution added. Here the end point corresponds to passage of the curve through the zero line. Representative curves plotted in these three ways are shown in Fig. 14.1.

The values of the first derivative may be established directly by a technique which is called a derivative potentiometric titration. Two identical electrodes are placed in the titration solution. One, however, is shielded by a glass tube with a narrow opening leading to the bulk of the solution (Fig. 14.2).

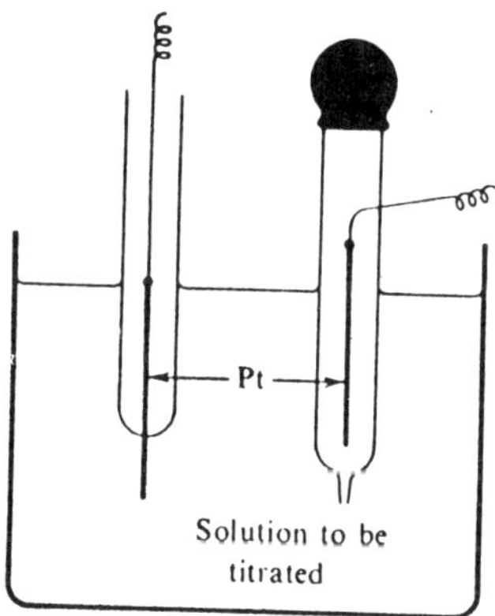

Fig. 14.2. Schematic diagram of an arrangement for a derivative potentiometric titration.

Initially both electrodes have the same potential and the cell voltage is zero. An increment of the titrant is then added and the solution is homogenized. The potential of the unprotected electrode changes because the composition of the solution has changed. However, the potential of the shielded electrode remains unchanged because it is still in contact with a solution of the composition present before the titrant addition. The voltage thus resulting is measured and recorded. Then the solution in the glass tube is flushed out by squeezing of the rubber bulb, and the tube is refilled from the bulk of the solution. Both electrodes are now again immersed in identical solutions; they are therefore at the same potential and the cell voltage is zero. This process is repeated for each (equal) increment of titrant. A graph of the cell voltage values, that if, of the potential differences versus the volume of titrant, corresponds to the first derivative of the usual titration curve. For equal increments the potential differences increase as the equivalence point is approached, reach a maximum when passing through this point, and become smaller again when an excess of titrant is added. At the equivalence point addition of the fixed amount of titrant causes the greatest fractional change in concentration and thereby in potential. Consequently, the maximum in the curve is taken as the end point.

Where a potentiometric titration is performed routinely, there is no need to plot the titration curve if the cell voltage corresponding to the end point is well established. The titrant solution need only be added until this voltage is reached. This modification of the potentiometric titration is readily capable of automation. Instruments are available that follow the potential and regulate the addition of titrant from a buret, the delivery of which is controlled electrically and stopped when the preselected cell voltage is reached. In other more sophisticated "autotitrators," a double differentiation is effected by an electronic circuit; when the output signal of the titrator is zero, a relay is triggered by means of which the flow of titrant is terminated. With this latter type of instrument, the potential of the equivalence point need not be established previously.

ELECTROLYTIC CONDUCTANCE AND CONDUCTOMETRIC TITRATIONS

Introduction

Electricity is conducted through a solution, that is, through an electrolytic conductor, by the motion of charged ions under the influence of an applied electrical field. The conductance is defined as the reciprocal of the resistance of the solution and is the summation of the contributions to the conductances of all ions in the solution. Each of these contributions is usually called an ionic conductance. The ionic conductance of an ion depends on its charge and the rate at which it migrates under the influence of the electrical field. Thus, a dipositive cation contributes twice as much to the conductance as a monopositive cation if both ions migrate at the same speed. The migration rate is influenced by

1. the magnitude of the applied electrical field,
2. the charge and size of the solvated ion,
3. the temperature,

4. the viscosity and dielectric properties of the solvent, and
5. the attractive forces between the ion of interest and other ions present in the solution.

In a concentrated solution, an ion is surrounded by other ions to which it is attracted and which retard its movement toward a charged electrode. If such a solution is progressively diluted, the attractive forces diminish and vanish at infinite dilution; thus, the ionic conductance increase with dilution and reaches a limiting value at infinite dilution. Frequently the approximation is made that the value of the ionic conductance at low finite concentrations (usually below 0.1 *M*) is equal to that at infinite dilution.

Units of Conductance

Conductance, as the reciprocal of resistance, is expressed as reciprocal ohms (abbreviated either mho or Ω^{-1}) and is directly proportional to the cross-sectional area (in square centimeters), *a*, and inversely proportional to the length (in centimeters), *l*, of a conductor.

$$\text{Conductance} = \frac{1}{R} = \frac{ka}{l} \quad \text{...(14.1)}$$

The proportionality constant, denoted by the Greek letter kappa, *k*, is called the specific conductance or, in the case of electrolytic conduction, the electrolytic conductivity, and has units of mhos per centimeter.

The equivalent conductance of an ion is designated by the Greek letter lambda, λ, and is defined as the electrolytic conductivity of a hypothetical solution that contains one equivalent of the ion per millilitre of solution. Since equivalents per millilitre equal equivalents per litre divided by 1000 it follows that

$$\lambda = \frac{k}{C^{*}/1000} \text{ mhos-cm}^2\text{/equivalent} \quad \text{...(14.2)}$$

C^{*} is the concentration in equivalents per litre and is here defined as the molarity multiplied by the absolute numerical value of the charge on the ion.

Solution of equation (14.2) for *k* and substitution in (14.1) yields

$$\text{conductance} = \frac{1}{R} = \frac{a\lambda C^{*}}{1 \times 1000} = K\lambda C^{*} \quad \text{...(14.3)}$$

where the proportionality constant *K* equals $a/(1 \times 1000)$. Each ion contributes independently to the conductance of the solution; hence,

$$\text{conductance} = \frac{1}{R} = K(\lambda_1 C_1^{*} + \lambda_2 C_2^{*} + \ldots)$$

$$= K \underset{1}{\Sigma} \lambda_1 C_1^{*} \quad \text{...(14.4)}$$

The equivalent conductance at infinite dilution is designed λ^0.

Application of Ohm's Law to Solutions

When a d.c. voltage is applied to two electrodes dipping into an electrolyte solution, Ohm's law is not obeyed. The causes for this fact have been considered in a previous section, and include the establishment of back emf, kinetic overvoltage, and concentration overvoltage.

If the polarity of the applied voltage is changed rapidly (that is, if an a.c. source is used) the situation is different. For a rather simplified picture, assume the polarity during the first half of a cycle is such that one of the electrodes is the cathode. Then cations are attracted to this electrode and anions are repelled; by the motion of the ions the electric current is carried through the solution. However, when an electrode reaction starts to take place, the second half of the cycle is reached and the polarity is reversed; that is, the electrode becomes the anode. Now cations are repelled and anions are attracted. Again before any appreciable reaction can occur the polarity changes, and so forth. Consequently, the ions carry the electrical current by a reciprocating motion, that is, merely by an oscillation, with no deposition taking place. Thus no back emf or overvoltage develops and Ohm's law is obeyed.

Measurement of Resistance

The measurement of the resistance and therefore the conductance of an electrolyte solution is usually performed with a modified Wheatstone

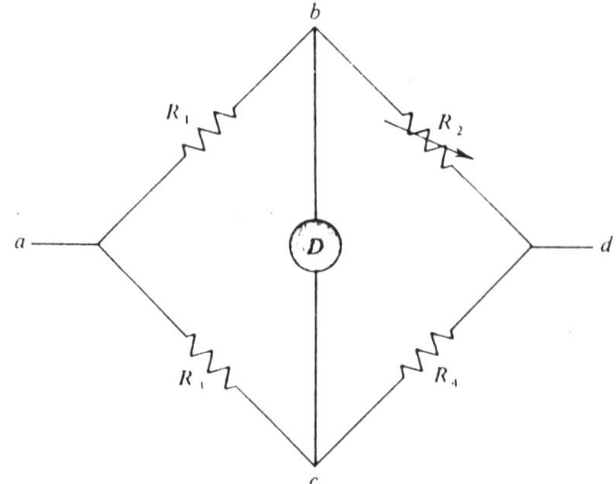

Fig. 14.3. Wheatstone bridge circuit. See the text for a detailed explanation.

bridge circuit, shown in Fig.14.3. A voltage is applied across points *a* and *d*. One (or more) of the resistors, R_1, R_2, R_3 and R_4, is variable and its resistance is accurately known at the various settings. By varying this resistance a setting can be found at which no current flows through the detector, *D*, and the bridge is said to be balanced.

Since no current passes through the detector when the bridge is balanced, points *b* and *c* must be at the same potential and consequently $I_1R_1 = I_3R_3$. Similarly, it follows that $I_2R_2 = I_4R_4$. Division of one equation by the other yields $I_1R_1/I_2R_2 = I_3R_3/I_4R_4$. Since no current passes through the detector, the current I_1 passing through R_1 must equal the current I_2 passing through R_2; that is , $I_1 = I_2$. Similarly, $I_3 = I_4$. Hence, for the balanced bridge, $R_1/R_2 = R_3/R_4$. If three of the resistances are known, the fourth is established and is thereby measured.

In the measurement of electrolytic conductance, it is customary to apply a pure sine-wave signal 10^2 to 10^3 cycles/sec) across points *a* and *d*.

Conductance Cells

A conductance cell often consists of a pair of platinized platinum electrodes rigidly mounted in a glass vessel. For any given cell, the values of *a* and *l* are fixed and the ratio *l/a* is called the cell constant, $K_{cell.}$ Although this constant might be calculated from the cell geometry, it is commonly evaluated by filling the cell with a solution of known electrolytic conductivity and measuring the resistance. The value of K_{cell} is then known from the relation $K_{cell} = kR$. Since conductance varies by about 2% per degree centigrade, it is necessary to control the cell temperature by use of a constant-temperature bath.

Conductivity Determinations

Conductivity determinations receive limited application in inorganic analysis. One of the most common uses is to check the quality of water intended for process or boiler-feed use. This measurement of conductivity is not specific for any particular impurity, but serves as a rough measure of the total ion content of the water. Carbon in steel and other metals and alloys is frequently determined by heating the sample in a tube furnace in a stream of oxygen. The exit gases are passed through a solution of an alkali metal or alkaline earth hydroxide. The decrease in the conductance of this solution can be correlated with the amount of carbon dioxide absorbed and this with the carbon content of the sample.

Conductometric Titrations

Any reaction in which the conductance of the reactants differs markedly from that of the products can, in principle, serve as the basis of a conductometric titration. In such a titration the conductance of the solution is measured after the addition of each increment of titrant solution and the values obtained are plotted versus the volume of titrant. The end point corresponds to a break in this curve and is located by the extrapolation of two linear segments. The conductance of the solution corresponds to the sum of the contributions of all ions present. Ions that do not enter into the titration reaction give a background conductance due to the titration reaction are superimposed. This background does not invalidate a conductometric titration unless it is so large that the small changes due to the titration are indiscernible. High concentrations of inert electrolytes are therefore undesirable in conductometric titrations.

Since the end point is obtained from *changes* in conductance, knowledge of the absolute value of the cell constant is not required and, in addition, the conductance may be expressed in arbitrary units. However, dilution effects must either be minimized or corrected for ; the temperature should be kept constant throughout the titration.

Conductometric Acid-Base Titrations

Consider the titration of hydrochloric acid with sodium hydroxide. The reaction equation for the present purposes may be expressed.

$$\underset{350}{(H^+} + \underset{76}{Cl^-)} + \underset{50}{(Na^+} + \underset{198}{OH^-)} \rightarrow H_2O + (Na^+ + Cl^-)$$

where the number below an ion is its λ^0 value.

The conductance associated with the hydrogen ion starts at a value proportional to its equivalent conductance (350) and decreases linearly to zero at the equivalence point. The conductance of the chloride ion remains constant throughout the titration at a value proportional to its equivalent conductance (76).

Hydroxide ion is not present in significant amounts until the equivalence point is passed, and then the conductance associated with this ion increases linearly to a value proportional to its equivalent conductance (198) when a 100% excess of titrant is added. The contribution to the conductance of the sodium ion starts at zero, is proportional to its equivalent conductance (50) at the equivalence point, and is twice that value (100) at the 100% excess point.

These relations are shown graphically as dashed lines in Fig.14.4

The point by point summation of these lines yields the theoretical titration curve (solid line). Only three points need be calculated to establish this curve: the starting point, equivalence point, and 100% excess point. This applies whenever the substances reacting are strong electrolytes. With weak electrolytes more points are needed and the calculations are involved.

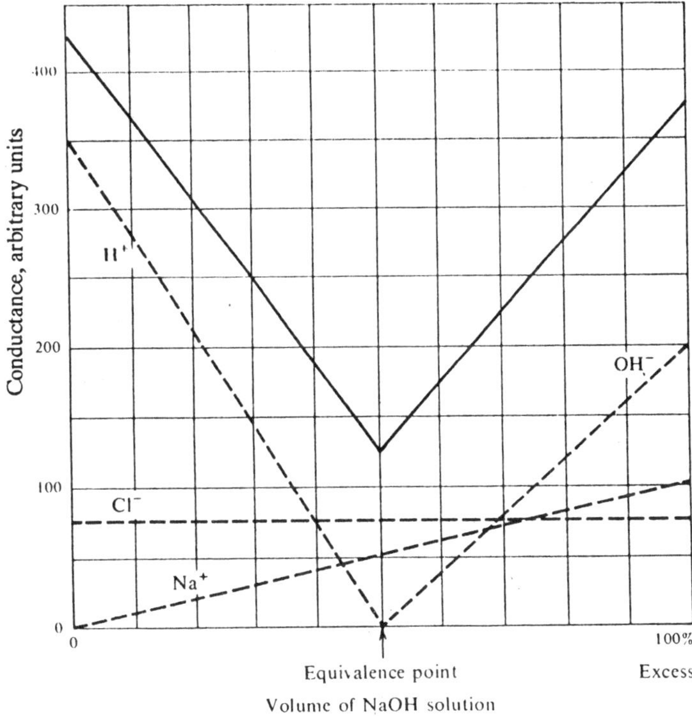

Fig. 14.4. Conductometric titration of hydrochloric acid with sodium hydroxide. The contribution of each ion to the conductance shown as a dashed line.

In the conductometric titration of weak acid, such as acetic acid, with sodium hydroxide (Fig.14.5, curve A), the conductance at the start is small, because only a small portion of the acid is dissociated. As sodium hydroxide is added, the conductance decrease at first because strongly conducting hydrogen ions are removed and acetate ions are formed that repress the dissociation of the acid and thereby prevent the formation of more hydrogen ion. As the titration proceeds the conductance increases nearly linearly up to the equivalence point. This increase is caused by introduction of conducting sodium ions by the titrant solution and the formation of acetate ions. Beyond that point, the conductance increases more rapidly with the addition of an excess of titrant and thereby introduction of highly conducting hydroxide ions. Since the slope of the titration curve changes only slightly at the equivalence point, it is difficult to locate the end point precisely. Analogous considerations apply to the titration of a weak base with a strong acid.

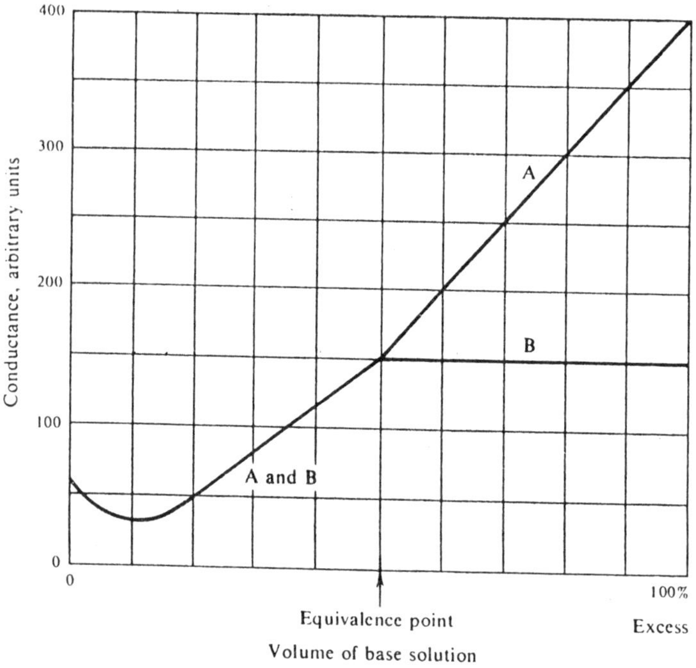

Fig.14.5. Curve A: conductometric titration of acetic acid with sodium hydroxide.
Curve B: conductometric titration of acetic acid with ammonia.

When a weak acid, such as acetic acid, is titrated conductometrically with a weak base, such as ammonia, the titration curve (Fig. 14.5, curve B) up to the equivalence point is similar to that for the titration of a weak acid with a strong base. Beyond that point, however, the conductance increases only slightly, since the ammonia is only partially dissociated. Further, its protolysis ($NH_3 + H_2O \rightleftharpoons NH_4^+ + OH^-$ is repressed by the ammonium ions present. Since the break in this curve is sharper than that with a strong base as titrant, the end point can be located for more precisely. This is quite different from the situation in a pH titration (potentiometric or visual) where a weak acid-weak base titration is infeasible.

Conductometric Precipitation Titrations

Consider the titration of sodium chloride with silver nitrate. The reaction equation for the present purposes may be expressed

$$\underset{50}{(Na^+} + \underset{76}{Cl^-)} + \underset{62}{(Ag^+} + \underset{71}{NO_3^-)} \rightarrow \underline{AgCl} + (Na^+NO_3^-)$$

where the number below an ion is its λ^0 value.

Since only strong electrolytes are involved, three points are sufficient to establish the shape of the theoretical titration curve. At the starting point, only sodium chloride is present and the conductance is proportional to 50 + 76 = 126. At the equivalence point, only sodium nitrate is present (and negligible amount of dissolved silver chloride) and the conductance is proportional to 50 + 71 = 121. At the 100% excess point, sodium nitrate and silver nitrate are present and the conductance is proportional to 50 + 71 + 62 + 71 = 254. The titration curve is shown in Fig.14.6.

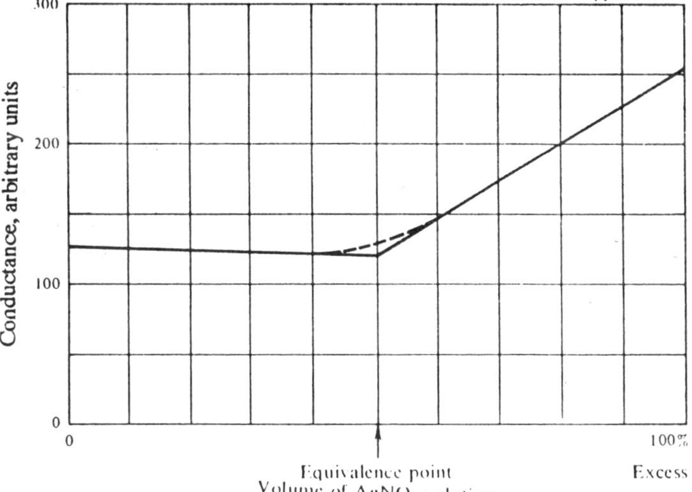

Fig.14.6. Conductometric titration of sodium chloride with silver nitrate.

If silver chloride were appreciably more soluble, then the conductance in the vicinity of the equivalence point would be increased measurably since the concentration of silver and chloride ions could not be neglected. In such a case, the two straight line-portions would be connected by a curved section (dashed line in Fig.14.6), but the end point could still be located precisely by extrapolation of straight-line segments.

Conductometric titrations can be applied to many chemical systems where an ionic reaction fails to go to completion. The dissociation of the principal product causes curvature in the titration plot in the region around the equivalence point: However, the dissociation is repressed before and beyond this point by a common-ion effect, thus yielding straight-line segments, which are extrapolated to establish the end point. In addition, conductometric titrations can be performed with dilute solutions, often 0.001 *F* or lower.

APPENDIX

TABLE 1 : SOME STANDARD AND FORMAL ELECTRODE POTENTIALS[a]

Half-Reaction	E° V	Formal Potential, V
$F_2(g) + 2H^+ + 2e \rightleftharpoons 2HF(aq)$	3.06	
$O_3(g) + 2H^+ + 2e \rightleftharpoons O_2(g) + H_2O$	2.07	
$S_2O_8^{2-} + 2e \rightleftharpoons 2SO_4^{2-}$	2.01	
$Co^{3+} + e \rightleftharpoons Co^{2+}$	1.842	
$H_2O_2 + 2H^+ + 2e \rightleftharpoons 2H_2O$	1.776	
$MnO_4^- + 4H^+ + 3e \rightleftharpoons MnO_2(s) + 2H_2O$	1.695	
$Ce^{4+} + e \rightleftharpoons Ce^{3+}$		1.70, 1-*F* $HClO_4$; 1.61 1-*F* HNO_3; 1.44, 1-*F* H_2SO_4
$HClO + H^+ + e \rightleftharpoons \frac{1}{2}Cl_2(g) + H_2O$	1.63	
$H_5IO_6 + H^+ + 2e \rightleftharpoons IO_3^- + 3H_2O$	1.60	
$BrO_3^- + 6H^+ + 5e \rightleftharpoons \frac{1}{2}Br_2(l) + 3H_2O$	1.52	
$MnO_4^- + 8H^+ + 5e \rightleftharpoons Mn^{2+} + 4H_2O$	1.51	
$Mn^{3+} + e \rightleftharpoons Mn^{2+}$		1.51, 7.5-*F* H_2SO_4
$ClO_3^- + 6H^+ + 5e \rightleftharpoons \frac{1}{2}Cl_2(g) + 3H_2O$	1.47	
$PbO_2(s) + 4H^+ + 2e \rightleftharpoons Pb^{2+} + 2H_2O$	1.455	
$Cl_2(g) + 2e \rightleftharpoons 2Cl^-$	1.359	
$Cr_2O_7^{2-} + 14H^+ + 6e \rightleftharpoons 2Cr^{3+} + +7H_2O$	1.33	
$Tl^{3+} + 2e \rightleftharpoons Tl^+$	1.25	0.17, 1-*F* HCl
$IO_3^- + 2Cl^- + 6H^+ + 4e \rightleftharpoons ICl_2^- + 3H_2O$	1.24	
$MnO_2(s) + 4H^+ + 2e \rightleftharpoons Mn^{2+} + 2H_2O$	1,23	1.24 , 1-*F* $HClO_4$
$O_2(g) + 4H^+ + 4e \rightleftharpoons 2H_2O$	1.229	

Table 1 (*Cont.*)

Half-Reaction	E° V	Formal Potential, V
$IO_3^- + 6H^+ + 5e \rightleftarrows \frac{1}{2} I_2(s) + 3H_2O$	1.195	
$IO_3^- + 6H^+ + 5e \rightleftarrows \frac{1}{2} I_2(s) + 3H_2O$	1.178[b]	
$SeO_4^{2-} + 4H^+ + 2e \rightleftarrows H_2SeO_3 + H_2O$	1.15	
$Br_2(1) + 2e \rightleftarrows 2Br^-$	1.065	1.05 4-*F* HCI
$Br_2(aq) + 2e \rightleftarrows 2Br^-$	1.087[b]	
$ICl_2^- + e \rightleftarrows \frac{1}{2} I_2(s) + 2Cl^-$	1.06	
$V(OH)_4^+ + 2H^+ + e \rightleftarrows VO^{2+} + 3H_2O$	1.00	1.02 -*F* HCI $HClO_4$
$HNO_2 + H^+ + e \rightleftarrows NO(g) + H_2O$	1.00	
$Pd^{2+} + 2e \rightleftarrows Pd(s)$	0.987	
$NO_3^- + 3H^+ + 2e \rightleftarrows HNO_2 + H_2O$	0.94	0.92, 1-*F* HNO_3
$2Hg^{2+} + 2e \rightleftarrows Hg_2^{2+}$	0.920	0.?07 1-F $HClO_4$
$HO_2^- + H_2O + 2e \rightleftarrows 3OH^-$	0.88	
$Cu^{2+} + I^- + e \rightleftarrows CuI(s)$ ·	0.86	
$Hg^{2+} + 2e \rightleftarrows Hg(I)$	0.854	
$Ag^- + e \rightleftarrows Ag(s)$	0.799	0.228 1-*F* HCI; 0.792 1-*F* $HClO_4$; 0.77; 1-*F* H_2SO_4
$Hg_2^{2+} + 2e \rightleftarrows 2Hg(I)$	0.789	0.274 1-*F* HCI; 0.776 1-*F* $HClO_4$; 0.674 1–*F* H_2SO_4
$Fe^{3+} + e \rightleftarrows Fe^{2+}$	0.771	0.700 1-*F* HCI; 0.732 1-*F* $HClO_4$; 0.68 1-*F* H_2SO^4
$H_2SeO_3 + 4H^+ + 4e \rightleftarrows 3H_2O$	0.740	
$PtCl_4^{2-} + 2e \rightleftarrows Pt(s) + Cl^-$	0.73	
$C_6H_4O_2$ (quinone) $+ 2H^+ + 2 \rightleftarrows C_6H_4(OH)_2$	0.699	0.696 1-*F* HCI, H_2SO_4 $HClO_4$
$O_2(g) + 2H^+ + 2e \rightleftarrows H_2O_2$	0.682	

Table 1 (*contd.*)

Half-Reaction	E° V	Formal Potential, V
$PtCl_6^{2-} + 2e \rightleftharpoons PtCl_4^{2-} + 2Cl^-$	0.68	
$Hg_2SO_4(s) + 2e \rightleftharpoons 2Hg(l) + SO_4^{2-}$	0.615	
$Sb_2O_5(s) + 6H^+ + 4e \rightleftharpoons 2SbO^+ + 3H_2O$	0.581	
$MnO_4^- + e \rightleftharpoons MnO_4^{2-}$	0.564	
$H_3AsO_4 + 2H^+ + 2e \rightleftharpoons H_3AsO_3 + H_2O$	0.559	0.577 1-*F* HCl $HClO_4$
$I_3^- + 2e \rightleftharpoons 3I^-$	0.536	
$I_2(s) + 2e \rightleftharpoons 2I^-$	0.5355	
$I_2(aq) + 2e \rightleftharpoons 2I^-$	0.620[b]	
$Cu^+ + e \rightleftharpoons Cu(s)$	0.521	
$H_2SO_3 + 4H^+ + 4e \rightleftharpoons S(s) + 3H_2O$	0.45	
$Ag_2CrO_4(s) + 2e \rightleftharpoons 2Ag(s) + CrO_4^{2-}$	0.446	
$VO^{2+} + 2H^+ + e \rightleftharpoons V^{3+} + H_2O$	0.361	
$Fe(CN)_6^{3-} + e \rightleftharpoons Fe(CN)_6^{4-}$	0.36	0.71 1-*F* HCl; 0.72 1-*F* $HClO_4$ H_2SO_4
$Cu^{2+} + 2e \rightleftharpoons Cu(s)$	0.337	
$UO_2^{2+} + 4H^+ + 2e \rightleftharpoons U^{4+} + 2H_2O$	0.334	
$BiO^+ + 2H^+3e \rightleftharpoons Bi(s) + H_2O$	0.32	
$Hg_2Cl_2(s) + 2e \rightleftharpoons 2Hg(l) + 2Cl^-$	0.268	0.242, satd KCl, 0.282 1-*F* KCl
$AgCl(s) + e \rightleftharpoons Ag(s) + Cl^-$	0.222	0.228 1-*F* KCl
$SO_4^{2-} + 4H^+ 2e \rightleftharpoons H_2SO_3 + H_2O$	0.17	
$BiCl_4^- + 3e \rightleftharpoons Bi(s) + 4Cl^-$	0.16	
$Sn^{4+} + 2e \rightleftharpoons Sn^{2+}$	0.154	0.14 1-*F* HCl
$Cu^{2+} + e \rightleftharpoons Cu^+$	0.153	

Table 1 (*contd.*)

Half-Reaction	E° V	Formal Potential, V
$S(s) + 2H^+ + 2e \rightleftarrows H_2S(g)$	0.141	
$TiO^{2+} + 2H^+ + e \rightleftarrows Ti^{3+} + H_2O$	0.1	0.04 1-*F* H_2SO_4
$AgBr(s) + e \rightleftarrows Ag(s) + Br^-$	0.095	
$S_4O_6^{2-} + 2e \rightleftarrows 2S_2O_3^{2-}$	0.08	
$Ag(S_2O_3)_2^{3-} + e \rightleftarrows Ag(s) + 2S_2O_3^{2-}$	0.01	
$2H^+ + 2e \rightleftarrows H_2(g)$	0.000	−0.005, 1-*F* HCl, $HClO_4$
$Pb^{2+} + 2e \rightleftarrows Pb(s)$	−0.126	−0.14, 1-*F* $HClO_4$;
		−0.29, 1-*F* H_2SO_4
$Sn^{2+} + 2e \rightleftarrows Sn(s)$	−0.136	−0.16, 1-*F* $HClO_4$
$AgI(s) - e \rightleftarrows Ag(s) + I^-$	−0.151	
$CuI(s) - e \rightleftarrows Cu(s) + I^-$	−0.185	
$N_2(g) - 5H^+ + 4e \rightleftarrows N_2H_{5+}$	−0.23	
$Ni^{2+} + 2e \rightleftarrows Ni(s)$	−0.250	
$V^{3+} + e \rightleftarrows V^{2+}$	−0.255	−0.21 1-*F* $HClO_4$
$Co^{2+} + 2e \rightleftarrows Co(s)$	−0.277	
$Ag(CN)_2^- + e \rightleftarrows Ag(s) + 2CN^-$	−0.31	
$Tl^+ + e \rightleftarrows Tl(s)$	−0.336	−0.551 1-*F* HCl;
$Cr^{3+} + e \rightleftarrows Cr^{2+}$		−0.33 1-*F* $HClO_4$ H_2SO_4
$PbSO_4(s) + 2e \rightleftarrows Pb(s) + SO_4^{2-}$	−0.356	
$Ti^{3+} + e \rightleftarrows Ti^{2+}$	−0.37	
$Cd^{2+} + 2e \rightleftarrows Cd(s)$	−0.403	
$Cr^{3+} + e \rightleftarrows Cr^{2+}$	−0.41	
$Fe^{2+} + 2e \rightleftarrows Fe(s)$	−0.440	

Table 1 (*contd.*)

Half-Reaction	$E°$ V	Formal Potential, V
$2CO_2(g) - 2H^+ + 2e \rightleftarrows H_2C_2O_4$	−0.49	
$Cr^{3+} + 3e \rightleftarrows Cr(s)$	−0.74	
$Zn^{2-} + 2e \rightleftarrows Zn(S)$	−0.763	
$Mn^{2+} + 2e \rightleftarrows Mn(s)$	−1.18	
$Al^{3+} + 3e \rightleftarrows Al(s)$	−1.66	
$Mg^{2+} + 2e \rightleftarrows Mg(s)$	−2.37	
$Na^+ + e \rightleftarrows Na(s)$	−2.714	
$Ca^{2+} + 2e \rightleftarrows Ca(s)$	−2.87	
$Ba^{2+} + 2e \rightleftarrows Ba(s)$	−2.90	
$K^+ + e \rightleftarrows K(s)$	−2.925	
$Li^+ + e \rightleftarrows Li(s)$	−3.045	

[a]Sources for E° values : W.M. Latimer The Oxidation States of the Elements and Their Potentials in Aqueous Solutions 2d ed. Englewood Cliffs N.J. : Prentice Hall Inc. 1952 ; A.J. de Bethune and N.A.S. Loud. Standard Aqueous Electrode Potentials and Temperature Coefficients at 25 C. Skokie, III. : Cliffara A. Hampel. 1964. Source of formal potentials B.H. Swift and E.A. Butler. Quantitative Measurements and Chemical Equilipria. W.H. Freeman and Company Copyright © 1972.

[b]These potentials are hypothecial since they correspond to solutions that are 1.00 *M* in Br_2 or I_2. The solubilities of these two compounds at 25°C are 0.21 *N* and 0.0133 *M* respectively. In saturated solutions containing an excess of Br_2 (I) or I_2 (s) the standard potentials for the half-reactions $Br_2(I) + 2e \rightleftarrows 2Br^-$ or $I_2(s) + 2e$ $2I^-$ should be used. On the other hand at Br_2 and I_2 concentrations less than saturation these hypothetical electrode potentials should be employed.

TABLE 2.

SOLUBILITY PRODUCT CONSTANTS

Substance	Formula	K_{sp}
Aluminum hydroxide	$Al(ON)_3$	2×10^{-32}
Barlum carbonate	$BaCo_3$	5.1×10^{9}

Table 2 (*contd.*)

Substance	Formula	K_{sp}
Barium chromate	$BaCrO_4$	1.2×10^{-10}
Barium iodate	$Ba(IO_3)_2$	1.57×10^{-9}
Barium manganate	$BaMno_4$	2.5×10^{-10}
Barium oxalate	BaC_2O_4	2.3×10^{-8}
Barium sulfate	$BaSO_4$	1.3×10^{-10}
Bismuth oxide chloride	BiOCI	7×10^{-9}
Bismuth oxide hydroxide	BiOOH	4×10^{-10}
Cadmium carbonate	$CdCO_3$	2.5×10^{-14}
Cadmium hydroxide	$Cd(OH)_2$	5.9×10^{-15}
Cadmium oxalate	$Cd C_2O_4$	9×10^{-8}
Cadmium sulfide	CdS	2×10^{-28}
Calcium carbonate	$CaCO_3$	4.8×10^{-9}
Calcium fluoride	CaF_2	4.9×10^{-11}
Calcium oxalate	CaC_2O_4	2.3×10^{-9}
Calcium sulfate	$CaSO_4$	1.2×10^{-6}
Copper (I) bromide	CuBr	5.2×10^{-9}
Copper (I) chloride	CuCI	1.2×10^{-6}
Copper (I) iodide	CuI	1.1×10^{-12}
Copper (I) thiocvanate	CuSCN	4.8×10^{-15}
Copper (II) hydroxide	$Cu(OH)_2$	1.6×10^{-19}
Copper (II) sulfide	CuS	6×10^{-36}
Iron (II) hydroxide	$Fe(OH)_2$	8×10^{-16}
Iron (II) sulfide	FeS	6×10^{-18}
Iron (III) hydroxide	$Fe(OH)_3$	4×10^{-38}
Lanthanum iodate	$La(IO_3)_3$	6.2×10^{-12}
Lead carbonate	$PbCO_3$	3.3×10^{-14}

Table 2 (*contd.*)

Substance	Formula	K_{sp}
Lead chloride	$PbCl_2$	1.6×10^{-5}
Lead chromate	$PbCrO_4$	1.8×10^{-14}
Lead hydroxide	$Pb(OH)_2$	2.5×10^{-16}
Lead iodide	PbI_2	7.1×10^{-9}
Lead oxalate	PbC_2O_4	4.8×10^{-10}
Lead sulfate	$PbSO_4$	1.6×10^{-8}
Lead sulfide	PbS	7×10^{-28}
Magnesium ammonium phosphate	$MgNH_4PO_4$	3×10^{-13}
Magnesium carbonate	$MgCO_3$	1×10^{-5}
Magnesium hydroxide	$Mg(OH)_2$	1.8×10^{-11}
Magnesium oxalate	MgC_2O_4	8.6×10^{-5}
Manganese (II) hydroxide	$Mn(OH)_2$	1.9×10^{-13}
Manganese (II) sulfide	MnS	3×10^{-13}
Mercury (I) bromide	Hg_2Br_2	5.8×10^{-23}
Mercury (I) chloride	Hg_2Cl_2	1.3×10^{-18}
Mercury (I) iodide	Hg_2I_2	4.5×10^{-29}
Silver arsenate	Ag_3AsO_4	1×10^{-22}
Silver bromide	$AgBr$	5.2×10^{-13}
Silver carbonate	Ag_2CO_3	8.1×10^{-12}
Silver chloride	$AgCl$	1.82×10^{-10}
Silver chromate	Ag_2CrO_4	1.1×10^{-12}
Silver cynide	$AgCN$	7.2×10^{-11}
Silver iodate	$AgIO_3$	3.0×10^{-8}
Silver iodide	AgI	8.3×10^{-17}

Table 2 (*contd,.*)

Substance	Formula	K_{sp}
Silver oxalate	$Ag_2C_2O_4$	3.5×10^{-11}
Silver sulfide	Ag_2S	6×10^{-50}
Silver thiocyhanate	AgSCN	1.1×10^{-12}
Strontium oxalate	SrC_2O_4	5.6×10^{-8}
Strontium sulfate	$SrSO_4$	3.2×10^{-7}
Thallium (I) chloride	TlCl	1.7×10^{-4}
Thallium (I) sulfide	Tl_2S	1×10^{-22}
Zinc hydroxide	$Zn(OH)_2$	1.2×10^{-17}
Zinc oxalate	ZnC_2O_4	7.5×10^{-9}
Zinc sulfide	ZnS	4.5×10^{-24}

TABLE 3

DISSOCIATION CONSTANTS FOR ACIDS

		Dissociation Constant 25°C		
Name	*Formula*	K_1	K_2	K_3
Acetic	CH_3COOH	1.75×10^{-5}		
Arsenic	H_3AsO_4	6.0×10^{-3}	1.05×10^{-7}	3.0×10^{-12}
Arsenious	H_3AsO_3	6.0×10^{-10}	3.0×10^{-14}	
Benzoic	C_6H_5COOH	6.14×10^{-5}		
Boric	H_3BO_3	5.83×10^{-10}		
1-Butanoic	$CH_3CH_2CH_2COOH$	1.51×10^{-5}		
Carbonic	H_2CO_3	4.45×10^{-7}	4.7×10^{-11}	
Chloroacetic	$ClCH_2COOH$	1.36×10^{-3}		
Citric	HOOC(OH) $C(CH_2COOH)_2$	7.45×10^{-4}	1.73×10^{-5}	4.02×10^{-7}
Ethylenediamine-tetraacetic	H_4Y	1.0×10^{-2}	2.1×10^{-2} $K_4 = 5.5$	6.9×10^{-7} $\times 10^{-11}$

Name	Formula	Dissociation Constant 25°C		
		K_1	K_2	K_3
Formic	HCOOH	1.77×10^{-4}		
Fumaric	*trans*-HOOCCH : CHCOOH	9.6×10^{-4}	4.1×10^{-5}	
Glycolic	$HOCH_2COOH$	1.48×10^{-4}		
Hydrazoic	HN_3	1.9×10^{-5}		
Hydrogen cyanide	HCN	2.1×10^{-9}		
Hydrogenfluoride	H_2F_2	7.2×10^{-4}		
Hydrogen peroxide	H_2O_2	2.7×10^{-12}		
Hydrogen sulfide	H_2S	5.7×10^{-8}	1.2×10^{-15}	
Hypochlorous	HOCl	3.0×10^{-8}		
Iodic	HIO_3	1.7×10^{-1}		
Lactic	$CH_3CHOHCOOH$	1.37×10^{-4}		
Maleic	*cis*-HOOCCH: CHOOH	1.20×10^{-2}	5.96×10^{-7}	
Malic	HOOCCHO HCH_2COOH	4.0×10^{-4}	8.9×10^{-6}	
Malonic	$HOOCCH_2COOH$	1.40×10^{-3}	2.01×10^{-6}	
Mandelic	$C_6H_5CHOHCOOH$	3.88×10^{-4}		
Nitrous	HNO_2	5.1×10^{-4}		
Oxalic	HOOCCOOH	5.36×10^{-2}	5.42×10^{-5}	
Periodic	H_5IO_6	2.4×10^{-2}	5.0×10^{-9}	
Phenol	C_6H_5OH	1.00×10^{-10}		
Phosphoric	H_3PO_4	7.11×10^{-3}	6.34×10^{-8}	4.2×10^{-13}
Phosphorous	H_3PO_3	1.00×10^{-2}	2.6×10^{-7}	
o-Phthalic	$C^6H_4(COOH)_2$	1.12×10^{-3}	3.91×10^{-6}	
Picric	$(NO_2)_3C_6H_2OH$	5.1×10^{-1}		
Propanoic	CH_3CH_2COOH	1.34×10^{-5}		
Pyruvic	$CH_3COCOOH$	3.24×10^{-3}		

Name	*Formula*	Dissociation Constant 25°C		
		K_1	K_2	K_3
Salicylic	$C_6H_4(OH)COOH$	1.05×10^{-3}		
Sulfamic	H_2NSO_3H	1.03×10^{-1}		
Sulfuric	H_2SO_4	strong	1.20×10^{-2}	
Sulfurous	H_2SO_3	1.72×10^{-2}	6.43×10^{-8}	
Succinic	$HOOCCH_2$ CH_2COOH	6.21×10^{-5}	2.32×10^{-6}	
Tartaric	$HOOC(CHOH)_2$ $COOH$	9.20×10^{-4}	4.31×10^{-5}	
Trichloroacetic	Cl_3CCOOH	1.29×10^{-1}		

TABLE 4

DISSOCIATION CONSTANTS FOR BASES

Name	Formula	Dissociation Constant
		K 25° C
Amonia	NH_3	1.76×10^{-5}
Aniline	$C_6H_5NH_2$	3.94×10^{-10}
1-Butylamine	$CH_3(CH_2)_2CH_2NH_2$	4.0×10^{-4}
Dimethylamine	$(CH_3)_2NH$	5.9×10^{-4}
Ethanolamine	$HOC_2H_4NH_2$	3.18×10^{-5}
Ethylamine	$CH_3CH_2NH_2$	4.28×10^{-4}
Ethylenediamine	$NH_2C_2H_4NH_2$	$K_1 = 8.5 \times 10^{-5}$ $K_2 = 7.1 \times 10^{-8}$
Hydrazine	H_2NNH_2	1.3×10^{-6}
Hydroxylamine	$HONH_2$	1.07×10^{-8}
Methylamine	CH_3NH_2	4.8×10^{-4}
Piperidine	C_5HN	1.3×10^{-3}
Pyridine	C_5H_5N	1.7×10^{-9}
Trimethylamine	$(CH_3)_3N$	6.25×10^{-5}